# Apple Cultivation - Recent Advances

*Edited by Ayzin Küden*

Published in London, United Kingdom

Apple Cultivation - Recent Advances
http://dx.doi.org/10.5772/intechopen.102140
Edited by Ayzin Küden

Contributors
Ranjna Sharma, Joginder Pal, Deepika Sharma, Satish Kumar Sharma, Shalini Verma, Radhika Pathania, Siti Khadijah A. Karim, Martina Komárková, Helena Cvrčková, Jaroslav Dostál, Václav Buriánek, Pavlína Máchová, Balnur Kabylbekova, Sagi Soltanbekov, Svetlana Dolgikh, Yuliya M. Yefremova, Marina V. Urazayeva, Saule Sh. Kazybayeva, Sweta Kumari, Swati Manohar, Priyanjali Kumari, Veda Krishnan, Chirag Maheshwari, Sneh Narwal, Om Prakash Gupta, Vinutha T. Gowda, Navita Bansal, Anil Dahuja, Richard M. M. Bastías, Alexandra Boini, Ayzin B. Küden, Ali Kuden, Burhanettin İmrak, Songül Comlekcioglu



First published in London, United Kingdom, 2023 by IntechOpen
IntechOpen is the global imprint of INTECHOPEN LIMITED, registered in England and Wales, registration number: 11086078, 5 Princes Gate Court, London, SW7 2QJ, United Kingdom

British Library Cataloguing-in-Publication Data
A catalogue record for this book is available from the British Library

Additional hard and PDF copies can be obtained from orders@intechopen.com

Apple Cultivation - Recent Advances
Edited by Ayzin Küden
p. cm.
Print ISBN 978-1-83969-800-2
Online ISBN 978-1-83969-801-9
eBook (PDF) ISBN 978-1-83969-802-6

# Meet the editor

Prof. Dr. Ayzin B. Küden is an academic staff member at the Faculty of Agriculture, Department of Horticulture, Çukurova University, Türkiye, where she served as the faculty's vice dean from 1997–1999 and dean from 2005–2012. She took part in the establishment of the ZIDEK Association with the assignment of the Agriculture, Forestry and Fisheries Deans Council in 2008, and still serves as the association's vice president. She chaired the TYYÇ Working Group on Agriculture, Forestry, Livestock and Fisheries, which was published in 2011. Dr. Küden's scientific interests include the conservation of genetic resources in fruit growing, physiological studies, stone fruit breeding, almond and fig selection and cultivation, the use of molecular markers in breeding, the physiology of dormancy in temperate climate fruits, and the effects of climate change on fruit growing. She has made short and long-term visits and training studies in many countries on these issues. Since 1996, she has been the chairperson of the Temperate Fruits in the Tropics and Subtropics (TFTS) Working Group of the International Society for Horticultural Science (ISHS). She is also a country member of ISHS. Dr. Küden is an editor and author with numerous publications to her credit. She is a journal referee and active on scientific committees. She has participated in six international symposiums, several workshops, national meetings, and national and international congresses. Although retired, Dr. Küden is still active in horticultural research and continues her duty as an executive in two EU Partnership on Research and Innovation in the Mediterranean Area (PRIMA) projects.

# Contents

# Preface

Cultivated apple (*Malus* x *domestica* Borkh.), which belongs to the *Rosaceae* family, is one of the most important fruit crops grown in the temperate zones of the world. Apples have been cultivated since ancient times. It is thought that the apple originated in Northern Anatolia, the Southern Caucasus, regions in the southwest of Russia, and around Central Asia. The wide adaptability of *Malus* allows it to be grown in many different climates. Despite high genetic variability, thousands of cultivars are distributed throughout the world. There are numerous studies on growing low-chilling apple cultivars in the subtropics and tropics. In some countries, apple production areas and rates are greater than those of other temperate fruit crops.

*Apple Cultivation – Recent Advances* discusses the many species of *Malus* in three sections: "Apple Germplasm, Molecular Characterization and *In Vitro* Culture", "Hormonal Regulation, Phytopathogens and Phenolics", and "Cultivars, Rootstocks, Nursery and Production Techniques of Malus Species".

**Dr. Ayzin Küden**
Professor,
Faculty of Agriculture,
Department of Horticulture,
University of Cukurova,
Adana, Türkiye

Section 1

# Apple Germplasm, Molecular Characterization and *In Vitro* Culture

Chapter 1

# Introductory Chapter: Apple Cultivation – Recent Advances

*Ali Kuden, Burhanettin Imrak, Songul Comlekcioglu and Ayzin Küden*

## 1. Introduction

Apple (*Malus* x *domestica* Borkh.) originated in Northern Anatolia, the Southern Caucasus, the regions in the southwest of Russia, and around Central Asia. This very nutritious fruit has spread all over the world from Central Asia. Apple has a wide adaptability to be grown in many climatic zones of the world. Besides Central Asia, East Asia, West Asia-Europe, and North America Gene Centers, countries such as Australia and New Zealand, which do not have gene centers, have played a major role in the emergence of new varieties.

The changes in apple production in European countries between 2014 and 2019 and the production estimates for 2020 are given in **Table 1**. While apple production decreased in Western European countries because of unfavorable weather conditions, Poland became the country with the highest increase in production (**Table 1**). When apple production in European countries between 2014 and 2019 was examined based on variety, there was a significant decrease in the production of Boskoop, Cox Orange, and Idared varieties grown with Golden Delicious and Red Delicious (Starking, Starkrimson, etc.) groups. Production increases that started with Jonagold and

| Countries | 2014 | 2015 | 2016 | 2017 | 2018 | 2019 | (1)2020 | (2)% |
|---|---|---|---|---|---|---|---|---|
| İtaly | 2.456 | 2.280 | 2.272 | 1.704 | 2.264 | 2.096 | 2.080 | 3 |
| France | 1.444 | 1.674 | 1.515 | 1.424 | 1.477 | 1.651 | 1.431 | −6 |
| Germany | 1.116 | 973 | 1.033 | 597 | 1.093 | 991 | 951 | 6 |
| Spain | 505 | 482 | 495 | 480 | 476 | 555 | 467 | −7 |
| England | 206 | 243 | 239 | 207 | 219 | 205 | 207 | −2 |
| Netherlands | 353 | 336 | 317 | 228 | 267 | 272 | 234 | −8 |
| Poland | 3.750 | 3.979 | 4.035 | 2.870 | 4.810 | 2.910 | 3.400 | −4 |
| Hungary | 920 | 522 | 498 | 530 | 782 | 452 | 350 | −40 |
| Romania | 382 | 336 | 327 | 230 | 425 | 327 | 343 | 5 |
| Total | 12.541 | 12.326 | 11.834 | 9.251 | 13.275 | 10.783 | 10.711 | −4 |

*(1) Product estimates. (2) Increases and decreases in production in 2019 and 2020.*

**Table 1.**
*Changes in apple production in Europe countries between 2014 and 2018 and 2020 production estimates.*

| Varieties | 2014 | 2015 | 2016 | 2017 | 2018 | 2019 | 2020 (1) | (2) |
|---|---|---|---|---|---|---|---|---|
| Cripps Pink | 249 | 244 | 261 | 260 | 275 | 289 | 277 | 1 |
| Elstar | 431 | 399 | 387 | 265 | 357 | 363 | 312 | −5 |
| Fuji | 321 | 338 | 288 | 290 | 332 | 316 | 294 | −6 |
| Gala | 1.318 | 1.382 | 1.312 | 1.271 | 1.467 | 1.439 | 1.490 | 7 |
| Granny S. | 383 | 405 | 384 | 363 | 393 | 372 | 369 | −2 |
| Jonagold | 644 | 633 | 567 | 298 | 577 | 391 | 310 | −27 |
| Jonagored | 491 | 519 | 539 | 335 | 563 | 246 | 231 | −39 |
| New Varieties | 165 | 207 | 213 | 209 | 344 | 359 | 405 | 33 |
| Pinova | 79 | 119 | 104 | 85 | 155 | 140 | 154 | 22 |
| Red Delici. | 675 | 643 | 632 | 558 | 737 | 678 | 660 | 0 |
| Red Prince | 98 | 104 | 156 | 114 | 371 | 407 | 437 | 47 |

*www.wapa-association.org & www.prognosfruit.eu*

**Table 2.**
*Changes in the production of apple varieties in Europe countries between 2014 and 2019 and the year 2020.*

Braeburn continued with Elstar, Gala, Fuji, and Pink Lady varieties. In recent years, the production of new varieties and club apples has started to increase, especially the Red Jona Prince (**Table 2**). Türkiye, Golden apple production in the previous years mostly depends on Golden Delicious, Starking Delicious, Granny Smith, and Amasya varieties and did not include many types and varieties in some statistics. With the popular use of new apple varieties and the establishment of a fully enclosed garden facility with them, statistics can be determined more easily. In Türkiye, the production of the apple varieties of Gala, Braeburn, Jonagold, Fuji, and Modi is estimated to be about 300,000 tons in 2020. This production amount constitutes a significant part of the total apple production. Apple production in the world apart from China, the varieties that are expected to increase in production are Cripps Pink, Honeycrisp, Gala, Scifresh / Jazz®, and Sciros / Pacific Rose®. Other varieties are significantly reduced [1, 2].

## 2. Pollination of apple

In addition to the varieties and rootstocks in the apple orchard establishment, other important factors are pollination and fertilization of the varieties. In the pollination process, biological and physical factors, sexual incompatibility, simultaneous flowering, pollinator vectors, and weather conditions at the time of pollination are important. When one or some of these factors are blocked or do not occur, pollination and consequently fruit set and quality are negatively affected. All apple varieties require foreign pollination in commercial cultivation. Self-productivity level varies according to varieties. For example, the Delicious variety is completely unfertile to itself, while Golden Delicious is partially fertile to itself. In terms of foreign pollination requests, all red and red spur types are similar to their parental plants and show incompatibility. These varieties can easily fertilize each other in the gardens that will include Gala apples, Fuji, Scarlet Spur, Redchief, and Granny Smith, which are also included in the experiments and are other commercial varieties. In general, the flowering period in apples lasts 10–15 days. Early and medium, and medium and late flowering varieties can pollinate each other. On the other hand, in addition to sufficient pollen production for pollination, flowering times must also coincide with each other.

For sufficient pollination;

1. Pollinator variety and main variety should bloom at the same time,

2. Pollinator varieties are diploid, and the vitality of flower pollens should be high,

3. The pollinator variety should not be too far from the main variety, it should be kept in the garden at appropriate rates,

4. Honeybee or other insect activities should be organized in the garden at the time of flowering,

5. Weeds, which attract more attention of bees and bloom at the same time with apples, should not be kept in the garden,.

6. In the selection of pollinator varieties, information such as the vitality of the pollen on the flowers of apple varieties and the flowering period of the main and pollinator variety are required.

Bee flight also becomes very weak in windy, rainy, and overcast weather. Therefore, wasps could be used for pollination in cold and rainy climates.

## 3. Apple rootstocks

### 3.1 Generative rootstocks

In many countries, seeds of Winesap, Rome Beauty, G. Delicious, Mclntosh, and Yellow Newton varieties are preferred to obtain rootstocks. In Türkiye, generative rootstocks of apples such as seeds of Golden Delicious and Ferik apple varieties can be advised to obtain rootstocks.

Although the high-density planting of apple orchards using dwarf and semidwarf rootstocks has gained importance in the world, apple orchards in Türkiye are still mostly established with apples grafted on seedlings.

Besides, deep-rooted and strong rootstocks are needed for the apple orchard plant in dry and insufficient irrigation areas.

### 3.2 Clonal apple rootstocks

Alnarp2 (A2): It was obtained in Sweden. It is a strong rootstock and suitable for cold regions.

AR Series: It was obtained at East Malling Horticultural Research Institute in England. The growing strength is between M9 and M27. It is better than M27 in terms of fruit size. The growing strength of the AR series range from M27 to MM106.

Budagovski Series: They are rootstocks obtained by crossing in Russia by Budagovski. The most important is B9. Other important Budagovski rootstocks are B146, B469, and B491.

CG Series: Selected from free pollinated seedlings of M8 by H. Guengerich in Stark Bros young plants in the USA. It is slightly stronger than the M9.

Cornell Geneva Series: It was obtained at New York Geneva Test Station. Geneva® breeding program is the first program to reveal several dwarf apple rootstocks (G41, G969, G214, G210, G202) that are more efficient and have a different resistance than MM106, and MM111 types have been.

J9 Series (Jork9): It was obtained from free-pollinated M9 seedlings in Germany. Some of its characteristics are similar to the M9.

JM Series: It was obtained by crossing medium-strong rootstocks Marubakaido (M.prunufolia) and Mitsubakaido (M.sieboldii) with M9 in Japan. The rootstocks that gain importance are JM1, JM5, JM7, and JM8. All of them are stunted from M26, while JM1, JM5, and JM8 are stunted from EMLA9.

KSC Series: Robusta No.5 main variety obtained in Kentvil in Canada and Antonovka rootstock were used as pollinators. The KSC is numbered from 1 to 30.

MAC Series: It was obtained by R.F Carlson at Michigan State University in the USA. MAC 1, 4, 9, 16, 24, 39, and 46 gained importance.

MH Series: British M and MM series rootstocks in Israel did not yield good results due to high soil temperatures and insufficient rest. MH 14–5 and MH 15–6 were obtained from Hashabi, a local apple variety.

Ottawa Series: It was obtained in Canada. It is named from 0.1 to 0.17. They are cold-resistant rootstocks.

P Series: These rootstocks are more resistant to cold than English rootstocks obtained by crossing Antonovka and M.4 and M9 in Poland. P1, P2, P16, P14, and P22 have gained importance.

V Series: It was obtained at Ontario Horticultural Research Station in Canada. Range from V.605–1 to V.605–7.

### 3.3 Series of East Malling very dwarf rootstocks

M27 was obtained by crossing M13 and M9 rootstocks in 1929. It is a very dwarf rootstock. It is not used to a great extent in practice. It is used as a transitional rootstock in strong rootstock weak-cutting relationships.

### 3.4 Dwarf rootstocks

M26 was obtained by crossing M16 and M9 rootstocks in 1959, and it has found a very fast usage area due to its earliness and 40–50% shortening compared to seedlings and productivity. It does not like very moist soils. It needs to be attached to a stake toward the ground to keep it in the ground.

M8, M9: The common characteristics of these two rootstocks are that the root systems are outcrop and fringe rooted; their holding in the soil is very weak. Also, trees grafted on both rootstocks begin to yield 2–3 years after planting, they become small-crowned and dwarfed. The dwarfing characteristics of M8 and M9 rootstocks are also used as transitional rootstocks. Besides, M9 rootstock dwarfs the apple trees by 65–75%.

### 3.5 Semidwarf rootstocks

Root depth on the soil of both M2 and M7 rootstocks is better than the previous rootstocks. They have medium-sized canopy and fructify early. These rootstocks are

| Varieties | Rootstocks | Planting distance (m) | Numbers of trees (pcs/ha) |
|---|---|---|---|
| Golden D. | M9 | 3,0 × 1,0 | 3333 |
| Granny Smith | M9 | 3,0 × 0,80 | 4166 |
| Jonagold | M9 | 3,0 × 1,0 | 3333 |
| For Red Delicious | | | |
| Standard types | M9 | 3,5 × 1,5 | 1904 |
| Standard types | MM 106 | 5,0 × 4,0 | 500 |
| Spur types | M 26 | 3,0 × 1,0 | 3333 |
| Spur types | MM 106 | 4,0 × 1,5 | 1666 |

**Table 3.**
*Rootstocks suitable for some apple varieties, planting spacing, and number of trees in one hectare.*

resistant to soil moisture. M2 and M7 rootstock are dwarfed by 25% and 35–45%, respectively, in apple varieties grafting on these rootstocks.

Strong rootstocks M1 and M13, M25, very strong rootstocks M12, M16. There are also Malling Merton rootstock series except for East Malling rootstocks. These are MM104, MM106, MM109, and MM111.

Among these rootstocks, MM106 is commonly used. The rootstock is semidwarfing and very fertile. It likes loamy soil, has a root system that can hold well in the soil, and tends to start very late in the spring development period, leaf fall, and winter rest period. MM 106 rootstocks induce 25–40% dwarfing in apple varieties grafting on the rootstock. The rootstock is not fertile in cropland with drainage problems or watery land.

The commonly used planting distances and the required seedlings for 10 decares considering the varieties and rootstocks are given in **Table 3**.

## 4. Club apples and varieties

Apple club varieties emerged in the late twentieth century and have been on the rise since the 2010s. In 2009, around 30 varieties of the apple club were distributed to stores around the world. The apple club continues mainly in Europe. Mostly, apple varieties play a role besides cherry and pear. In this system, the world market is managed by centralized hiring, and apples are produced in limited quantities under license and brand protection in high standard quality. Prospective producers negotiate with the licensee about production and marketing before joining the club.

Producers are required to ensure the fruit quality criteria determined in the fields of breeding and marketing. With quality control, fruits are sold at a high price or at least a standard price is provided. Production is controlled and prevented by nonmembers.

Major important club apple varieties: Civni-Rubens®, FavDiwa® (Junami®), Pink Lady®, Rosy Glow, Modi®, Kanzi, Ambrosia, Jazz™, Lady Alice, Pacific Rose™, Piñata®, Sonya, Sweetango™, Red Prince, Caudle (Cameo®), Cripps Pink (Pink Lady®), Delblush, (Tenta- tion®), Honeycrisp (Honeycrunch®, Brak (Kiku® 8. Nicogreen-(Greenstar®), Sciros, Pacific Rose®.

## 5. Importance of apple for health

Apple is the most consumed fruit in the world and is important for health. Apple is a powerful source of antioxidants and it has been determined to increase resistance against some types of cancer, cardiovascular diseases, asthma, and diabetes. In studies, it has been determined that apples contain strong antioxidants, such as catechin, floridzin, and chlorogenic acid, various chemical compounds and quercetin, thus reducing lipid oxidation and reducing cholesterol and preventing the proliferation of cancer cells. It is stated that there are 1500 mg of vitamin C in 100gr fresh apples, and that polyphenols, quercetin, and glycosides are found in the apple peel. It has been determined that apples contain two times more antioxidants than citrus fruits, about five times more than bananas, and 10 times more than Goji berry juice [3].

## 6. Covering systems in apple orchards

Nets are mostly used for protection from rain, winter frosts, and sun. As in the rest of the world, increases in the value of sunlight receiving our country have been observed in the last 5 years. In some ecologies in the temperate climate zone where apple cultivation is carried out, it is known that apples are sensitive to high temperature and solar radiation (light intensity), which is seen especially in summer and causes sunburn in fruits; however, it varies according to apple types. In apple cultivation, sunburn damage occurs on fruits at temperatures above 35–40°C and this damage negatively affect the quality and storage time of the fruit. While sunburn damage occurs in fruits with high light intensity and lux, fruits are not colored well with low light intensity [4].

## 7. Pruning of apple trees

Pruning in apple orchards are made in three groups:

a. Important points in shape pruning on young trees: Over-pruned seedlings bear fruit late, so the seedlings should be pruned according to the purpose. Branch angles should be considered while creating side branches. Therefore, the variety and characteristics of the rootstock should be well known. The shape of the apple trees could be created depending on the region [5].

b. The purpose of pruning on productive trees is to keep the crown of the tree at a certain size and to ensure trees' yields every year.

c. Rejuvenation pruning is applied to trees after 30–35 years of age to ensure fruit size and economical production. Since rejuvenation pruning is very severe, pruning should be completed in 2–3 years and technical works, such as fertilization, irrigation, and agricultural struggle, should be done carefully. Grafting paste should be used after the pruning processes. In the dwarf system, renewal pruning should be applied to keep the trees young.

In fruit trees, there are three separate life cycles: the period between planting and the beginning of the tree to yield is called "Juvenil," the period from yield to losing productivity is called "maturity," and the following period is called "old age."

Farmers want the period of juvenility to last as short as possible. Therefore, they use some technical measures to get shorted this period, such as cutting roots, drowning the stem, fertilizing fruit seedlings with nitrogen fertilizers in a balanced way, and using weak rootstocks. In addition, creating a regular, strong, and balanced canopy in fruit trees, leaving branches long or not cutting at all, widening the angles of strongly growing branches or narrowing the angles of poorly growing branches, cutting off some of the excess branches, and bending or twisting branches are some used methods to balance between root and crown in trees in a shorter time.

### 7.1 Pruning times

*Winter pruning:* The period after fruit trees enter winter dormancy is the best time for pruning in places where the winter is warm. In regions with cold winters, pruning is done after severe frosts to prevent the fruit trees from getting cold on the cutting surfaces, and the resistance of the tree against frost is increased. The most suitable period for pruning fruit trees is the period between the days following the defoliation and the beginning of the spring development period.

*Summer pruning:* All of the processes, such as reducing the shoots from brunches during the summer, taking the tip, bending, connecting the branches and widening, and narrowing the angles, are called summer pruning. When the branches of fruit trees begin to become woody, summer pruning should begin and continue this process in June, July, and August, and continue this process until the end of September in some cases. After that, winter pruning should not be done. Summer pruning is an important technical procedure that should be done especially during the shaping years of fruit trees. On the other hand, the branches that need to be removed by cutting can be used by bending, twisting and such branches, which are harmful to trees, can be used well. Therefore, early initiation of the flower bud of fruit trees can be achieved and their yields can be increased.

*Yield pruning:* Fruit trees enter physiological balance at the end of certain periods. The balance between shoot and fruit development should continue for a long time. This can only be achieved by yield pruning.

*Branch break:* In modern apple cultivation, when there is an imbalance of development on side branches in trees, usually backward pruning is performed in Spur and Standard apple varieties. Recently, the branch-breaking method is performed when there is an unbalance of development in the side branches in standard varieties.

### 7.2 Tree shape pruning in apples trees

*Classic orchards (Free Shapes):* Goble, central leader, modified leader.

using modern dwarf rootstocks and compulsory shapes: Palmets, Cordons (Vertical and Inclined Cordon), Spindel, Super Spindel, Tall Spindel, Normal Spindel, Vertical Axixs, HYTEC, Solen, Solax, Mikado, and Bibaum.

*Shape criteria in pruning, and relationship between row planting spaces and trunk diameter.*

Canopy projection diameter less than 50 $cm^2$ on Cords and row distances 40–50 cm.

Canopy projection diameter 50–100 cm on Super Spindel, Fruit Wall, Bibaum, High Tree Shape, row distances 50–100 cm.

Canopy projection diameter greater than 100 cm Normal Spindel, Solax, Tall Spindel, Bibaum, row distances 100–120 cm.

In classic dwarf orchards, tree heights were limited to 2.2–2.5 meters in spindle pruning. In these orchards, as the trees get older, the pruning labor cost increases with each passing year. The figure below shows trees that have not been pruned and pruned on classic dwarf apples. Too many cuts are applied here.

## 8. Thinning methods in apple

Thinning in apple trees is especially important in terms of fruit quality. The purpose of thinning is to increase the amount of marketable fruits.

Nowadays, fruit thinning is done by hand, hormones, and chemicals. Hand thinning in 15–20 cm per fruit or one fruit per 40 to 50 leaves, in the spindle system, pruning is done at 20–25 fruits per tree approximately 30 days after full bloom.

In the thinning studies carried out on apples, chemical substances, and hormones applications provide regular yields every year, while the hand thinning applications decrease the yield in the following year.

## 9. Soil properties and soil cultivation of apple trees

The best soil for apple growing is sufficient amount of loam, loamy-sandy, and sandy-loam permeable soils, including lime and humus. Apple trees are more sensitive in terms of soil characteristics in an arid area than in a moist area.

Although apple roots generally grow upward, the groundwater is not desired to be above 1 meter. If there is less salt in the soil, the tree is able to grow in this situation. The apple tree is more sensitive to unfavorable soil conditions than the pear tree.

Trees in very calcareous soils turn yellow due to problems with iron absorption. The most favorable soil reaction is between pH 6 and 8 levels. Soil tillage methods to be applied in apple orchards able to be formed into two main parts.

*Open soil tillage:* The purpose is to keep clean the bottom of the trees and the orchards by using all kinds of tillage tools in summer and winter. Thus, it is possible to eliminate the competition of weeds with plant roots. The open tillage method should never be applied in a field that is widely exposed to water and wind erosion.

*Conservation soil tillage:* As the name suggests, this method is to keep the soil surface covered with some cover plants. In this system, depending on whether the cover is permanent or not, it is divided into two temporary covered and permanent covered.

Autumn cover crops, such as wheat, rye, winter vetch, red truffle, sweet truffle, and clover, are planted in temporary coverings. In the spring, the cover plants are buried in the soil by using a disc harrow.

*Mulching:* It provides benefits in numerous fields, such as preserving moisture, weed control, temperature, reducing labor and cost, protects the soil structure, improving soil structure, reducing nutrient and water loss, high-quality fruit, pest and disease control, early harvesting, and high-quality yield.

## 10. Conclusion

Recently, modern apple-growing techniques are developed and spread all around the apple-growing areas. With the fast development in science and technology, understanding the importance of apple on human health, increase and easiness of transportation and exportation, gained more importance to apple growing. Thus, in different countries in the world, apple's marketing and growing techniques developed by the results of many research studies and publications.

*Malus domestica* Borkh. is one of the most important fruit species widely spread to various climatic and soil conditions with the use of different rootstocks. This wide adaptability of Malus species gives the opportunity to be grown in many different climatic areas. Despite high genetic variability, thousands of cultivars are distributed throughout the world. There are also many studies on growing apples in the subtropics and tropics with low-chilling apple cultivars.

## Author details

Ali Kuden, Burhanettin Imrak, Songul Comlekcioglu and Ayzin Küden*
Faculty of Agriculture, Department of Horticulture, University of Cukurova, Adana, Türkiye

*Address all correspondence to: abkuden@gmail.com

## References

[1] Binard P. The World Apple and Pear Association. 2020. www.wapa-association.org & www.prognosfruit.eu

[2] Küden A, Küden AB, Imrak B, Sarier A. Apple growing in Turkey. Acta Hortic. 2019;**1261**:25-28. DOI: 10.17660/ActaHortic.2019.1261.5

[3] Eberhardt M, Lee C, Liu R. Antioxidant activity of fresh apples. Nature. 2000;**405**:903-904. DOI: 10.1038/35016151

[4] Kuden A, İmrak B. Apple growing in TURKEY. In: New Approaches in Apple and Cherry Growing and Breeding Techniques (Work Shop). Turkey, Adana; 2014. pp. 37-41

[5] Barritt BH. Management of HYTEC apple orc- hards. Part II: Pruning and training. Good Fruit Grower. 1992;**43**(19):38-42

Chapter 2

# *Malus* Wild Species of Kazakhstan and Their Conservation *In Situ*

*Svetlana Dolgikh, Sagi Soltanbekov and Balnur Kabylbekova*

## Abstract

Kazakhstan has concentrated unique genetic resources of plant agrobiodiversity of world significance. It has received international recognition for fruit agrobiodiversity and, above all, wild apple, which is highly resistant to many diseases, high frost resistance, and wide ecological plasticity. Kazakhstan is the original genetic center of biological variability of wild apples and, historically, has formed their rich gene pool. Wild apple species in Kazakhstan are genetically kindred to cultivated varieties of the world. Genetic diversity and polymorphism of *Malus Sieversii* (Ledeb.) from Ile Alatau and Dzhungraian Alatau were studied with ISSR markers.

**Keywords:** wild apple, *Malus Sieversii*, genetic diversity, in situ, conservation

## 1. Introduction

The *wild apple tree* of Kazakhstan is represented by *M. Sieversii*, *M. Kirghisorum*, and *M. Niedzwetzkyana*. There are quite serious disagreements regarding the systematic position of these species. So F.D. Likhonos [1] believes that according to the morphological characteristics of the Nedzvetsky apple tree—*M. Niedzwetzkyana* is very close to the Sievers apple tree—*M. Sieversii* and is defined as a variety of *M. Sieversii* subsp. *Sieversii* var. *Niedzwetzkyana* (Dieck) Likh. An undeniable fact is that the Nedzvetsky apple tree is distributed within the area of the Sievers apple tree, but according to V.T. Langenfeld [2], the presence of only red pigmentation of vegetative parts, flowers, and fruits of the plant does not allow considering this apple tree as an independent species. The apple tree of the Kyrgyz is *M. Kirghisorum*, which does not have large morphological differences from the apple tree of Sievers V.T. Langenfeld, proposes to consider as a subspecies *M. Sieversii* subsp. *Kirghisorum* (Theod.et.Fed.) Likh. In modern conditions, these species, according to the peculiarities of addition and distribution, by their valuable role and share of participation in the vegetation cover, perform to various degrees. Thus, *M. Sieversii* occupies a wider and defined range, while *M. Kirghisorum's* role in the formation of the modern landscape looks secondary and topographically limited. *M. Niedzwetzkyana*, discovered by B.A. Bykov in 1957 in Karatau and Ile Alatau, is insignificant in number and range and does not form large populations anywhere.

 IntechOpen

So, the natural forests of wild apple of the Kazakhstan in terms of scale, uniqueness, genetic potential, scientific, and practical significance can be classified as one of the most valuable plant communities of the earth [3]. In this regard, their study and preservation *in situ* and *ex situ* are important and relevant for the scientific community.

## 2. Specific features of *Malus* wild species of Kazakhstan and implication for their conservation

### 2.1 Specific features of *Malus* wild species of Kazakhstan

Sievers' apple tree is 2–10 (14) m tall. The leaves are large from short-elliptic to oblong with a wedge-shaped, less often rounded base, at the apex usually suddenly turning into a short pointed tip, along the edge the leaf plate is town-shaped. The flowers are 3.5–6 cm in diameter, on long felt pedicels. The petals are pale pink. Fruits are 3–4 cm in diameter, spherical, or flattened-spherical. Along mountainous slopes, gorge bottoms, river valleys in the belt of woody-shrubby vegetation (from 900 to 2300–2600 m above sea level) form apple forests.

*M. Sieversii* is a very polymorphic species. Experimental data from population genetics show that in plants in the mountains, due to the small size of populations and the variety of natural conditions, random combinations of features that are little controlled by selection prevail. This leads to the emergence of new qualities and the rapid pace of their evolution, which was the main reason for the emergence in the mountainous regions of Central Asia of a huge variety of forms and polymorphism of apple trees. In mountain forests, wild individuals are distinguished by an amazing variety of fruits. These apple trees are drought-tolerant, frost-tolerant, and durable (about 150 years) [4].

Wild apple hybrids are characterized by a huge intraspecific polymorphism as a consequence of genetic differences. Within one population, each apple tree is an independent form with reliably distinctive features and properties from the neighboring apple tree. The practical value of these forms and their hereditary properties vary greatly; therefore, in order to preserve the most valuable and well adapted to the specific environmental conditions of the forms, it is necessary to be planted with planting material genetically corresponding to the selected forms. Vegetative propagation meets this condition. *Wild apple tree* refers to a culture that is easily propagated by root growth, but difficult to form subordinate roots on the stem parts. Non-specialized parts and organs of the plant that serve the purpose of vegetative propagation are not naturally separated from the mother plant. On horizontally located lateral roots of apple trees, adventitious (appendage) and dormant buds develop, from which, under favorable conditions, root offspring form. *M. Kirghisorum and M. Sieversii* are distinguished by great energy in the formation of root offspring and large natural stands, numbering up to several hundred trunks, turn out to be connected by a common root system. And among them is a viable mother tree. The formation of root offspring occurs throughout the horizontal roots at a depth of 30–40 cm. Numerous offspring arise at a distance of 15–20 m from the tree and grow in rows, sometimes up to 500 pieces in one tree [5], forming apple groves, having a common root system, and genetically being one vegetatively propagated specimen. It has been established that vegetative apple trees do not reduce growth processes and do not show signs of aging trees.

Recently, the *wild apple tree* of Kazakhstan has been represented by depleted populations in habitats, greatly destroyed by the established practice of nature management. The area of these forests in 2000 was only 7% compared with 1930 [5].

Therefore, along with the problem of conservation, there is a global problem of restoring wild apple *in situ* and creating *ex situ* collections. *In situ*, biodiversity conservation is the conservation of ecosystems and natural habitats, as well as the maintenance and restoration of viable species populations in their natural environment. The *ex situ* method of preservation and restoration includes methods such as cryopreservation, field gene banks (uteri of root and seed material), propagation, and preservation *in vitro*.

The general provisions of the theory of conservation of species through their reintroduction have been repeatedly discussed in the literature [6]; however, practical work with a specific species requires the development of a special strategy due to the specific reasons that cause the "rarity" of the species. Most researchers indicate that when reintroducing a species, it is necessary to follow a number of rules to achieve genetic and ecological prosperity of populations mainly in the former habitats of the species [7]. To do this, the mother material must be represented by the most heterozygous plants from the populations with the highest polymorphism indicators. An important step is the identification of parent and daughter plants and their compliance with reintroduction sites [8, 9].

Efficient conservation and use of plant genetic resources require a thorough assessment of the genetic variation they possess.

Genetic variation can be measured at two levels:

- Phenotype—combinations of individual traits determined by the genotype and its interaction with the environment.

- Genotype—specific genetic structure of the organism.

The reintroduction strategy for biodiversity recovery should involve recovery of population-typical allele frequencies. Historically, all patterns of variability and heredity have been studied by analyzing morphological characteristics, which until now have remained the main criterion in plant taxonomy, in the study of the mutation process, in studies of phylogenetic connections. However, morphological features do not always provide complete informatively, since they are strongly influenced by environmental conditions. In this regard, along with morphological features, physiological and biochemical indicators are widely used, for example, polymorphism of isoenzymes and spare proteins; however, isoenzymes and proteins are products of gene expression, the degree of expression of which is influenced by various factors, for example, plant age and tissue specificity [10]. If we consider that in higher eukaryotes only about 1% of the genome are protein-coding sequences, then the main part of the genome escapes the attention of researchers.

Currently, methods based on polymerase valuable reaction (PCR) using molecular markers characterizing regions of variability in nuclear, chloroplast, and ribosomal DNA are widely used worldwide in taxonomy and population genetics.

Various evolutionary events result in different variants in the DNA sequence that describes the polymorphism. Polymorphism manifests itself in genotype differences and is visualized as different band profiles found when using molecular markers in PCR and electrophoresis of PCR products.

The most simple and preferable in experiments where DNA of a large number of representatives at several loci is studied are methods using DNA amplification by single primers with an arbitrary sequence of SSR [11] and ISSR nucleotides [12].

ISSR uses known microsatellite sequences as primers and carries at one end a sequence of two to four arbitrary nucleotides. Such primers allow amplification of DNA fragments that are located between two sufficiently close microsatellite sequences (usually unique DNA). The obtained PCR product patterns are species-specific [13]. ISSR markers refer to markers of dominant inheritance type whose polymorphism is tested by the presence/absence of a band. The method has good reproducibility and, along with AFLP, is used to identify interspecies and intra-specific genetic variation, identify species, populations, lines, and individual genotyping [14].

### 2.2 Molecular genetic assessment of *M. Sieversii* in Kazakhstan[1]

With ISSR markers, polymorphism and genetic diversity of *M. Sieversii* in the Ile Alatau and Dzungarian Alatau populations have been studied [15].

Plant material *of M. Sieversii* was selected from the populations of Ile Alatau and Dzungar Alatau on expeditions in 2009–2011 organized by UN in Kazakhstan. Samples of young leaves without signs of bacterial and fungal infections taken closer to the point of growth of shoots were fixed in silica gel. Genomic DNA from 400 samples *M. Sieversii* (Ile Alatau and the Dzungarian Ile Alatau) was extracted from dry leaves (20 mg of each sample) using whales of NucleoSpin Plant (-Macherey-Negel, Germany) according to the protocol of the producer and was stored at t -25°C.

*M. Sieversii* inter-simple sequence repeats (ISSRs) were amplified using primers (**Table 1**) on a Mastercyeler ep gradient amplifier (Eppendorf). The primers used were synthesized at the Russian company Synthol (**Table 1**). All polymerase chain reactions (PCRs) took place in a total volume of 20 μl.

In the course of work on the selection of ISSR markers, polymorphic between the studied genotypes for all markers, the same PCR conditions were used (primer annealing was set individually), allowing to obtain the maximum amount of reaction product.

| № primer | Nucleotide sequence (5′-3′) | № primer | Nucleotide sequence (5′-3′) |
|---|---|---|---|
| M2 | aca cac aca cac aca c(ct)g | M8 | gtg gtg gtg gtg gtg |
| M3 | gag aga gag aga gag a(ct)c | M9 | gac acg aca cga cac gac ac |
| M4 | aga gag aga gag aga g(ct)c | M11 | cac aca cac aca (ag) |
| M7 | cag cag cag cag cag | M12 | cac aca cac aca (ag)(ct) |

**Table 1.**
*Primer nucleotide sequence.*

[1] Studies were carried out within the framework of the Project of the Government of the Republic of Kazakhstan, GEF/UNDP "Conservation in situ of mountain agro-biodiversity in Kazakhstan."

PCR parameters used for analysis included: 3 minutes at 95°C—initial denaturation, the following 35 cycles:

30 seconds denaturation at 94°C, annealing primers at the appropriate temperature—30 seconds,

40 seconds elongation at 72°C + addition of 2 seconds per cycle.

The reaction mixture (20 μl) contained 10–20 ng of DNA, 20 pmoles of primer and a finished reaction mix (Biocom) containing Taq-inhibited DNA polymerase, deoxynucleoside triphosphates, and magnesium chloride with final concentrations of 1u, 200 μM, and 2.5 mM, respectively, as well as an optimized buffer system for PCR.

Initially, the annealing temperature of primers was calculated using the formula:

$$T = 4°C \text{ x } (C + G) + 2°C \text{ x } (A + T) - 3, \quad (1)$$

where C, G, A, and T are the amount of cytosine, guanidine, adenine, and thymine bases, respectively. Subsequently, the optimal annealing temperature of the primers was empirically selected by decreasing or increasing it, depending on the quality of the PCR product obtained. The primer annealing temperature was specifically for primers M11 to 44.8°C, M12 to 49.5°C, M2 to 49.5°C, M9 to 50°C, M8 to 52.7°C, M3 to 52.7°C, M4 to 50.8°C, and M7 to 52.7°C.

For electrophoretic separation of PCR products, 1.7% agarose gel in 1x tris-borate buffer (50 mM Tris, 50 nM boric acid, 1 mM EDTA, pH 8.0) with ethidium bromide (0.5 μg/mL) was used when 100V for 45 minutes, followed by photographing the obtained PCR products and photo-processing in adobe program. Photographs of agarose gels were analyzed in Cross Checker 2.91 [13] with the composition of binary arrays of the presence/absence of fragments of the same length.

The level of heterozygosity of marker data is identified according to the average similarity frequency of alternative alleles. At the same time, 50% of the main alleles versus 50% of alternative alleles (0.5) correspond to a high level of heterozygosity, while 0.9 by 0.1 (90% versus 10%) corresponds to a low level.

**Figures 1–5** show electropherograms of amplification of DNA fragments of *M. Sieversii* samples obtained with three primers M8, M3, and M2. ГД—Golden Delicious cultivar, AP—Aport cultivar.

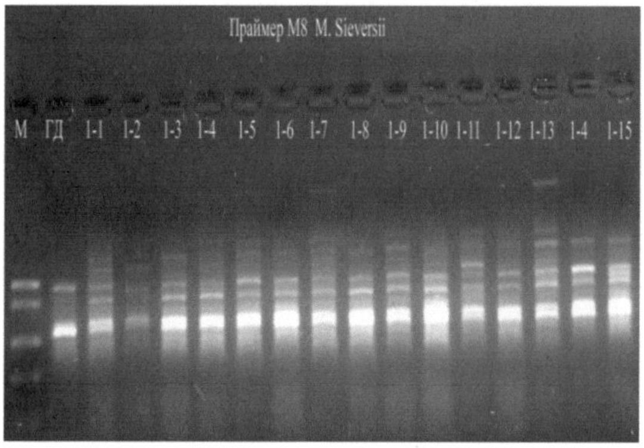


**Figure 1.**
*Electropherogram of amplification products of apple tree DNA fragments obtained by M8 primer.*

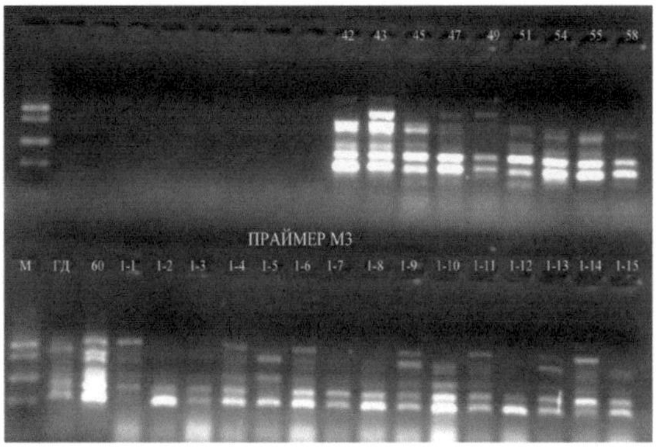


**Figure 2.**
*Electropherogram of amplification products of apple tree DNA fragments obtained by M3 primer.*

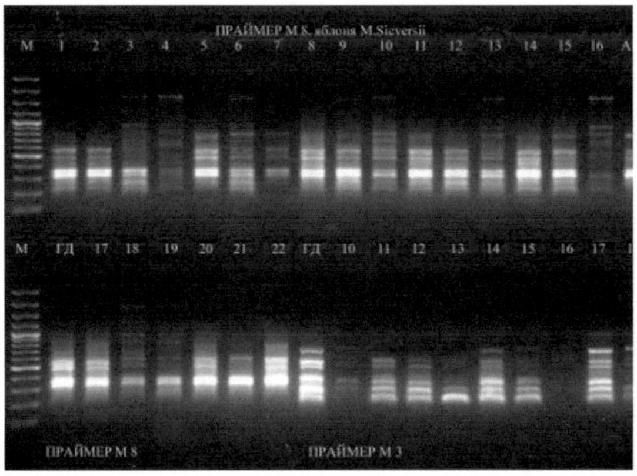


**Figure 3.**
*Electropherogram of apple tree DNA fragment amplification products obtained using primers M8 and M3.*

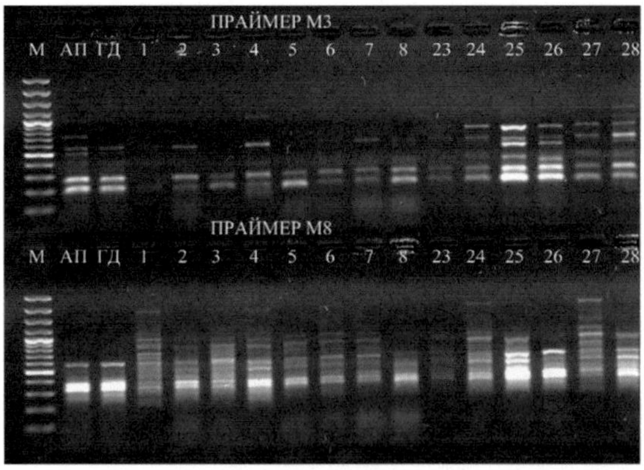


**Figure 4.**
*Electropherogram of apple tree DNA fragment amplification products obtained using primers M8 and M3.*

**Table 2** shows the genetic diversity of *M. Sieversii* from the Ile Alatau population, and **Table 3** shows the genetic diversity of *M. Sieversii* from the Dzungar Alatau population, which was calculated as the percentage of alternative amplicon ISSR profiles relative to the overall ISSR profiles of *M. domestica* amplicons.

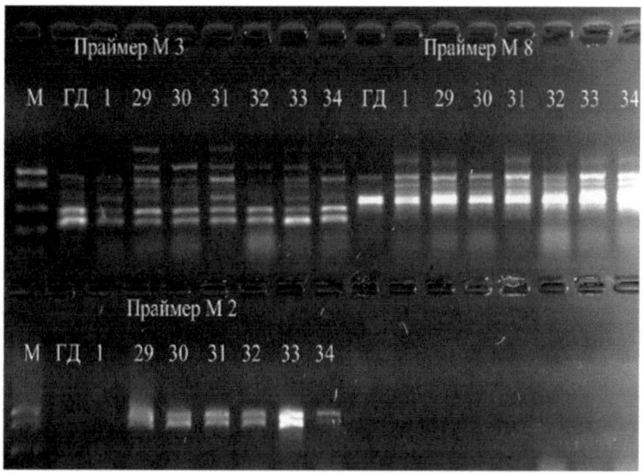


**Figure 5.**
*Electropherogram of apple tree DNA fragment amplification products obtained using primers M2, M8 and M3.*

| Sample no. | Common amplicons with cultural apple, % | Genetic diversity, % | Sample no. | Common amplicons with cultural apple, % | Genetic diversity, % |
|---|---|---|---|---|---|
| 1-1 | 40 | 60 | 23 | 40 | 60 |
| 1-2 | 50 | 50 | 29, 32, 33 | 50 | 50 |
| 1-3 | 50 | 50 | 37 | 67 | 33 |
| 1-4 | 33 | 67 | 41, 44 | 50 | 50 |
| 1-5 | 28,5 | 71,5 | 42 | 50 | 50 |
| 1-6, 1-15 | 50 | 50 | 43 | 43 | 57 |
| 1-7, 1-13 | 33 | 67 | 45 | 50 | 50 |
| 1-8, 1-9, 1-14 | 50 | 50 | 47 | 75 | 25 |
| 1-10 | 50 | 50 | 48, 49, 52 | 33 | 67 |
| 1-11 | 40 | 60 | 51 | 60 | 40 |
| 1-12 | 50 | 50 | 53 | 66 | 34 |
| 1 | 50 | 50 | 54, 55 | 50 | 50 |
| 4,7 | 67 | 33 | 56 | 40 | 60 |
| 12, 16, 27, 40 | 40 | 60 | 57 | 78 | 22 |
| AP, GD, 13, 36 | 100 | – | 58 | 67 | 33 |
| 18,19 | 50 | 50 | 60 | 50 | 50 |
| 21 | 50 | 50 | | | |

**Table 2.**
*Genetic diversity of M. Sieversii (Ile Alatau).*

As can be seen from **Tables 4–8**, the Golden *Delicious* and Aport cultivars and wild forms have the same ISSR profiles containing several amplicons, which characterizes their relation to the genus *Malus* and various amplicons showing polymorphism of the species. Overall, the polymorphism of *M. Sieversii* and in Ile Alatau and Dzungar

| Sample no. | Common amplicons with cultural apple, % | Genetic diversity, % | Sample no. | Common amplicons with cultural apple, % | Genetic diversity, % |
|---|---|---|---|---|---|
| 1 | 93 | 7 | 18 | 50 | 50 |
| 2 | 78 | 22 | 19 | 30 | 70 |
| 3 | 45 | 55 | 20 | 50 | 50 |
| 4 | 30 | 70 | 21 | 50 | 50 |
| 5 | 74 | 26 | 22 | 60 | 40 |
| 6 | 70 | 30 | 23 | 50 | 50 |
| 7 | 85 | 15 | 24 | 33 | 67 |
| 8 | 80 | 20 | 25 | 60 | 40 |
| 9 | 56 | 44 | 26 | 90 | 10 |
| 10 | 43 | 57 | 27 | 36 | 64 |
| 11 | 47 | 53 | 28 | 43 | 57 |
| 12 | 43 | 57 | 29 | 46 | 54 |
| 13 | 50 | 50 | 30 | 45 | 55 |
| 14 | 45 | 55 | 31 | 42 | 58 |
| 15 | 43 | 57 | 32 | 44 | 56 |
| 16 | 26 | 74 | 33 | 46 | 54 |
| 17 | 80 | 20 | 34 | 42 | 58 |

**Table 3.**
*Genetic diversity of M. Sieversii (Dzungar Alatau).*

| Sample no. | Common amplicons with cultural apple, % | Genetic diversity, % | Sample no. | Common amplicons with cultural apple, % | Genetic diversity, % |
|---|---|---|---|---|---|
| 50, 56, 57, 58, 59, 61, 62, 63, 83, 89, 92: 97 | 100 | | 79; 81; 82 | 40 | 60 |
| 52; 53 | 49 | 51 | 86; 87 | 35 | 65 |
| 79; 81; 82 | 30 | 70 | 98; 99; 100; 101; 103 | 50 | 50 |
| 64; 65 | 45 | 55 | 51; 55; 60; 68; 69; 70; 71; 73; 74; 76; 83; 84; 88; 90; 93; 94; 95; 102; 104 | | 100 |
| 78; 80 | 43 | 57 | | | |

**Table 4.**
*Genetic diversity of M. Sieversii (Dzungar Alatau).*

Alatau was 73%, with genetic diversity in the Ile Alatau population being higher at 82% and in Dzungar Alatau at 70%. At the same time, in Ile Alatau, 60% of samples were grouped with *M. domestica*, and in Dzungar Alatau—25%.

| Sample no. | Common amplicons with cultural apple, % | Genetic diversity, % | Sample no. | Common amplicons with cultural apple, % | Genetic diversity, % |
|---|---|---|---|---|---|
| ChR-17; ChR-21; ChR-22; ChR-11; ChR-20 | 100 | | KR-1; KR-2 | 45 | 55 |
| ChR-2; ChR-12 and ChR-23 | 48 | 52 | KR-18; KR-4; KR-22 | 40 | 60 |
| ChR-16; ChR-24 | 49 | 51 | SR-3-4 | 100 | |
| ChR-9; ChR-10 | 50 | 50 | SR-3-6; SR-3-11 | 49 | 51 |
| ChR-25; ChR-3; ChR-8; ChR-18; ChR-7; ChR-6; ChR-1; ChR-14; ChR-15; ChR-5 | | 100 | SR-3-13; SR-3-3 | 45 | 55 |
| KR-19; KR-24; KR-14; KR-23; KR-26 | 100 | | SR-2; SR-3-11; SR-3-9; SR-3-7; SR-3-8; SR-3-10; SR-3-5 | | 100 |
| KR-5; KR-3; KR-6; KR-21; KR-12; KR-13; KR-10; KR-11; KR-25; KR-27; KR-7; KR-15; KR-28; KR-8; KR-20; KR-18 | | 100 | | | |

**Table 5.**
*Genetic diversity of Malus Sieversii (Dzungar Alatau).*

| Sample no. | Common amplicons with cultural apple, % | Genetic diversity, % | Sample no. | Common amplicons with cultural apple, % | Genetic diversity, % |
|---|---|---|---|---|---|
| Ssh -2 №7 and Ssh -2 №9 | 49 | 51 | AP, PS-17; PS-7; PS-12 and PS-18 | 100 | |
| Ssh -2 №4 кв9; Ssh -1-2; Ssh -1-3 | | 100 | 105, 106 and 107 (Kuznetsova Gorge) | 45 | 55 |
| DA KD –2 №17 and DA KD–2 №18; DA KD–2 №16 and DA KD–2 №19; DA Ssh 1-5; DA Ssh 1-2 and DA Ssh 1-1; | 40 | 60 | PS-8; PS-9; PS-14; PS-15; PS-11; PS-10; 108 and 109 (Kuznetsova Gorge); DA KD–2 №14; DA KD–2 №15; DA Ssh 1-3 and DA Ssh 1-4 | | 100 |

**Table 6.**
*Genetic diversity of M. Sieversii (Dzungar Alatau).*

| Sample no. | Common amplicons with cultural apple, % | Genetic diversity, % | Sample no. | Common amplicons with cultural apple, % | Genetic diversity, % |
|---|---|---|---|---|---|
| KD 1-8; 1-13; 1-14; 1-15; 1-16 | 35 | 65 | 2-№19; 2-№20 | 35 | 65 |
| KD 1-2;1-3; 1-4; 1-12; 1-16; 1-6; 1-11; 1-1;1-7; 1-9 and 1-10 | | 100 | 2-№4; 2- №5; 2- №9 | | 100 |
| 2-№11; 2-№12 | 40 | 60 | AP, DA Ssh 1-6; Ssh 1-7; Ssh 1-8; Ssh 1-9; Ssh 1-10; Ssh 1-15; Ssh 1-16; Ssh 1-17 | 100 | |
| 2-№1; 2-№6; 2-№7; 2-№8; 2-№10; 2-№13 | 33 | 67 | Ssh 1-20; Ssh 1-21 | 50 | 50 |
| 2-№2; 2-№3 | 35 | 65 | Ssh 1-12; Ssh 1-14; Ssh 1-18; Ssh 1-19 | | 100 |
| 2-№14; 2-№15; 2-№16 | 45 | 55 | | | |

**Table 7**
*Genetic diversity of M. Sieversii (Dzungar Alatau).*

| Sample no. | Common amplicons with cultural apple, % | Genetic diversity, % | Sample no. | Common amplicons with cultural apple, % | Genetic diversity, % |
|---|---|---|---|---|---|
| KZh-22; KZh-18; KZh-38; KZh-40 | 100 | | KZh-43 and KZh-44 | 45 | 55 |
| KZh-23; KZh-25 and KZh-34 | 49 | 51 | D-9; гз-14 and GZ-15 | 40 | 60 |
| KZh-20 and KZh-21 | 45 | 65 | USh-4 and USh -5 | 35 | 65 |
| KZh-32; KZh-33 and KZh-46 | 49 | 51 | DA Ssh -2 №9 and DA Ssh -2 №11 | 49 | 51 |
| KZh-45; KZh-47; KZh-48 and KZh-49 | 49 | 51 | KZh-19; KZh-24; KZh-26; KZh-27; KZh-28; KZh-29; KZh-30; KZh-31; KZh-35; KZh-36; KZh-37; KZh-39; KZh-41; KZh-42; KZh-50; D-7; D-8; D-10; D-11; D-12; GZ-13; GZ-16; GZ-17; KT-1; KT-2; KT-3; UB-6; DA Ssh -2 №2; DA Ssh - №5 | | 100 |
| 51x, 52x and 54x | 50 | 50 | 53x | | 100 |

**Table 8.**
*Genetic diversity of M. Sieversii (Dzungar Alatau).*

*DOI: http://dx.doi.org/10.5772/intechopen.109401*

## 3. Conclusion

Thus, through molecular genetic analysis of 400 wild apple samples from the Ile Alatau and Dzungarian Alatau populations, ISSR markers have established the presence of loci that can serve as markers for identifying species and populations and identifying intraspecific genetic variation. To create living collections of *ex situ* and reintroduction of *M. Sieversii*. 22% of *M. Sieversii* forms from Ile Alatau and 63% of *M. Sieversii* forms from Dzungar Alatau, which amounted to, in total, 170 samples of wild apple. The theory of conservation of plant species of reintroduction provides for the admissibility of introducing into a population only those genotypes that are historically present in it. Otherwise, the genetic structure of the population is disturbed, which can lead to irreversible consequences. Therefore, the restoration of degraded plantations should be carried out with seedlings containing the genotypes of these particular plantations, and at the same time free of cultivated fruit genes.

## Acknowledgements

Studies were carried out within the framework of the Project of the Government of the Republic of Kazakhstan, GEF/UNDP "Conservation *in situ* of mountain agro-biodiversity in Kazakhstan." Project partially funded and published under the scientific program BR10765032 of the Ministry of Agriculture of the Republic of Kazakhstan.

## Author details

Svetlana Dolgikh, Sagi Soltanbekov and Balnur Kabylbekova*
Kazakh Fruit and Vegetables Research Institute, Almaty, Kazakhstan

*Address all correspondence to: kabylbekova.balnur@gmail.com

## References


[1] Likhonos FD. Semechkovs. Cultural Flora. 1983;**14-S**:16-125

[2] Langenfeld VT. Morphological evolution. Geography. 1991;**1991**:233

[3] Aleksanyan SM, Sedov EN, Gorbunov YN, Dolgikh SG, Maxted N, Rauzin EG, et al. Modern Methods and International Experience in Preserving the Gene Pool of Wild Plants (Using the Example of Wild Fruit Plants). Almaty, Kazakhstan. 2011. p. 210

[4] Dzhangaliev AD. Wild apple tree of Kazakhstan. Alma-Ata. 1977;**1977**:281

[5] Dzhangaliev AD. The wild apple tree of Kazakhstan. Horticultural Reviews. 2003;**29**:63-303

[6] Falk DA. Restoring Diversity: Strategies for Reintroduction of Endangered Plants. Washington, Covelo: Island Press; 1996. p. 505

[7] Schemske DW, Husband BC, et al. Evaluationg approaches to the conservation of rare and endangered plants. Ecology. 1994;**1994**:384-606

[8] Dolgikh SG, Mishenko AB. Molecular-genetic estimation of intraspecific diversity of Malus Sieversii and Armeniaca Vulgaris in Kazakhstan. Current Opinion in Biotechnology. 2011; **22**:820

[9] Dolgikh SG, Rauzin EG, et al. Recommendations for the Preservation of Clones in Archives (Live Collections). Almaty, Kazakhstan. p. 40

[10] Hosaka K, Matsubayasi M, Kamijima O. Peroxidase isozyme in various tissues for discrimination of two tuberose Solanum species. Japan Journal of Breeding. 1985;**35**:375-382

[11] Volk GM, Richards CM, Reilley AA, Henk AD, Forsline PL, Aldwinckle HS. *Ex situ* conservation of vegetatively-propagated species: Development of a seed-based core collection for *Malus Sieversii*. Journal of American Society. 2005;**130**:203-210

[12] Bornet B, Branchard M. Nonanchored Inter Simple Sequence Repeat (ISSR) Markers: Reproducible and specific tools for genome fingerprinting. Plant Molecular Biology Reporter. 2001;**19**:209-215

[13] Camacho FJ, Liston A. ISSR population analysis of the Oregon endemic, Botrychium pumicola (Ophioglossaceae). American Journal of Botany. 1998;**85**(6):99-100

[14] Ge XJ, Sun M. Reproductive biology and genetic diversity of a cryptoviviparous mangrove Aegiceras corniculatum (Myrsinaceae) using allozyme and intersimple sequence repeat (ISSR) analysis. Molecular Ecology. 1999;**8**(12):2061-2069

[15] Dolgikh SG. Molecular genetic assessment of intraspecific diversity of wild-growing crops in Kazakhstan. In: International Scientific and Practical Conference: Wild-Growing Forests of Kazakhstan: Issues of Conservation and Rational Use of the Gene Pool of Global Significance. Almaty, Kazakhstan. 2012. pp. 24-27

**Chapter 3**

# Long-Term Observation of *In Vitro*-Derived Malus Sylvestris (L.) Mill., the Path from the Bud to the Tree

*Martina Komárková, Helena Cvrčková, Jaroslav Dostál, Václav Buriánek and Pavlína Máchová*

## Abstract

The European wild apple *(Malus sylvestris* L.), a wild contributor to the domesticated apple, belongs to the endangered species in the Czech Republic. Thus, an efficient protocol was developed for *in vitro* plantlet regeneration using the post-dormant buds. The highest shoot induction was obtained on MS medium supplemented with 0.5 $mg.l^{-1}$ BAP, 5 $mg.l^{-1}$ GA and 0.1 $mg.l^{-1}$ IBA. Shoot multiplication and elongation took place on the same medium with 0.2 $mg.l^{-1}$ BAP and 0.1 $mg.l^{-1}$ IBA. Indole-3-butyric acid at 0.5 mg. $l^{-1}$ was most effective for rooting. The micropropagated plantlets were successfully acclimatized in greenhouse conditions and were transplanted into soil in forest. Finally, qualitative and quantitative parameters of tissue culture-derived plants were evaluated. Monitoring of *in vitro* plantings on experimental trials suggests that micropropagated wild apple trees retain the growth characteristics of generative individuals.

**Keywords:** acclimatization, establishment of *in vitro* culture, long-term forest trial, *Malus sylvestris* L., rooting, shoot multiplication

## 1. Introduction

*M. sylvestris* L., the only native wild apple species in Europe, belongs to the family Rosaceae. It is mainly pollinated by bees and flies, and thanks to its small and hard fruits, the trees are often called crab apples. This species occurs across Western and Central Europe, from southern Scandinavia to the Iberian Peninsula and from the Volga to the British Isles [1]. With a high up to 10 m, it grows in low-density populations and the growth habit looks like shrubs, more than trees. Under good growth conditions (high light requirements), the crab apple can live up to 80–100 years [2]. Though many different varieties of apples were developed over time, it was shown that old varieties have higher nutritional quality when compared to commercial ones [3–5].

In the Czech Republic, the wild apple belongs to the endangered species [6] because of its shrinking habitat, fragmentation of populations [7–9] and likelihood of a genetic admixture with domesticated apple (Malus domestica Borkh.). Recent studies have underlined the significant contribution of wild apples to the cultivated apple genome, M. domestica Borkh., during the domestication history from the Central Asian progenitor *Malus sieversii* (Ldb.) M. Roem [10]. Because of hybridization, it can be very difficult to reliably identify 'true type' wild apples from existing hybrids based on morphological characteristics, such as the fruit width or the hairiness of leaves [2]. Easy hybridization of wild apple with the domesticated apple is due to the absence of prezygotic isolation mechanisms [11], which results in fitness reduction of the wild apple populations and can lead to their reduction or extinction. Moreover, due to human activities, such as forest clearing, industrialization, population increase and intensive agriculture, natural and artificial forest stand regeneration is almost impossible. To avoid loss of wild apple genetic resources, they must be conserved and the conservation strategy should be applied sustainably.

The solution to the problem of preservation and restoration of the gene pool of endangered wild apples could be the *in vitro* cultivation using the micropropagation method. Micropropagation is used to multiply a wide variety of plants by a number of tissues, cells or organ culture methods. It means the aseptic culture of small plant explant of tissue or organs, closed in vessel with defined culture media and under controlled conditions. It provides the large-scale production of disease-free seedlings within a short time and with limited space. In the Czech Republic, biotechnology approaches for *in vitro* conservation of different woody plant species are well established and widely used (e.g., see [12–17]). During the past years, many studies have been carried out using *in vitro* cultures of apples (for review, see [18]). A key objective of apple *in vitro* cultures is to multiply disease-free clones, suitable for rooting, acclimatization and planting. However, as in many other plant species, the medium composition, plant growth regulators requirement or specific growth conditions are cultivar-dependent [19].

Additionally, micropropagation enables the protection of endangered species, thanks to providing the plant material in a larger amount for plant breeding programs at specific sites [20]. However, not many available papers and reviews dealing with studies on tissue cultures of forest trees include subsequent *ex vitro* evaluation of their quantitative and qualitative traits. The aim of our study was to find the optimal composition of explant culture media used for initiation of culture, as well as long-term multiplication and rooting of wild apples *in vitro*. In addition, the growth parameters of *in vitro*-derived wild apple trees were evaluated.

## 2. Material and methods

### 2.1 Plant material and *in vitro* culture establishment

*M. sylvestris* L. trees were carefully determined, not to be mistaken for domesticated apples M. domestica Borkh.

Juvenile branches with dormant leaf buds of the crown of M. sylvestris trees from central Bohemia were collected and cut into 10–15 cm long twigs. Twigs were immediately packed into the plastic bags and stored at 4°C before being cultured *in vitro*. To decrease microbial contamination, the twigs were cut into 5 cm long segments with buds, rinsed for 30 min in running water and surface disinfected with Tween®20 (2 drops/10 ml) for 20 min, followed by soaking in KORSOLEX for 20 min, rinsed with sterile distilled water for 20 min, incubated in $HgCl_2$ solution (1 $mg.l^{-1}$) for 15 min and subsequently immersed in distilled water three times for 15 min.

The surface disinfected bud explants (0.3–0.5 cm large) were placed in glass jars containing 100 ml of culture medium: MS medium [21] supplemented with 0.5 mg $l^{-1}$ BAP, 0.1 mg $l^{-1}$ IBA, 10 mg $l^{-1}$ glutamine, 2 mg $l^{-1}$ glycine, 30 g $l^{-1}$ sucrose and 6 g $l^{-1}$ agar. The pH value was adjusted to 5.8 before autoclaving at 121°C, 150 kPa for 20 min. The aseptic cultures were incubated in a growth chamber at 24 ± 1°C with a 16-h photoperiod (30 μE $m^{-2}$ $s^{-1}$).

### 2.2 Shoot multiplication

New shoots from explant cultures were separated into stems approximately 2.0 cm long and transferred into shoot induction media. The medium used was MS medium containing 0.2 mg $l^{-1}$ BAP, 0.1 mg $l^{-1}$ IBA, 200 mg $l^{-1}$ glutamine, 2 mg $l^{-1}$ glycine, 200 mg $l^{-1}$ casein, 30 g $l^{-1}$ sucrose, 6 g $l^{-1}$ agar and final pH adjusted to 5.8. Explants were cultured in the growth conditions as described above and repeatedly subcultured at a constant 4-week subculture interval.

### 2.3 Root induction, *ex vitro* acclimatization and hardening

Healthy shoots (1.5–2 cm) were excised and cultured on ¼ MS medium supplemented with 0.5 mg $l^{-1}$ IBA, 10 g $l^{-1}$ sucrose, 6 g $l^{-1}$ agar and final pH adjusted to 5.8. Explants were cultured in the growth conditions as described above.

After 2–3 weeks, the rooted shoots were carefully washed with distilled water to remove agar and the rooted shoots were hardened in a culture room in conical planter Quick Pot T 35 with perforated bottom, filled with perlite (Perlite Praha spol. s.r.o., Czech Republic) and placed into the transparent plastic box fully closed. Plants were regularly watered with 1/2 strength liquid MS medium devoid of sucrose and phytohormones and diluted with distilled water in a ration 1:10. After 2–3 weeks, the plants were transferred into the planter Quick Pot T 60 containing soil (Zahradnický substrát a.s. Soběslav, Czech Republic): perlite (2:1) and kept in a bigger transparent plastic box. During 2 weeks, plantlets were adapted to lower moisture conditions by gradually tilting the upper part of the plastic box. Fully adapted plants were moved into greenhouse and subsequently transferred into the outdoor flower beds to obtain capable seedlings.

## 3. Results

### 3.1 Plant material

The wild apple trees are characterized by pressed buds, leaves and combs. Only in spring, the leaves can have inconspicuous hairs on the reverse side at the base of the thicker veins. The size of the leaves and fruits are also different, with clearly smaller size in the wild apple tree. *M. sylvestris* L. has leaves of up to 6.5 cm long, while the leaves of the domestic apple tree are 6–12 cm long. The fruits (pome) of the wild apple tree are 2–3.5 cm in diameter, while the apples of the domestic apple tree are at least 5 cm in size. The crown slices of flowers of the wild apple tree are also smaller and tend to be narrower [22].

Juvenile branches with axillary buds were collected during the early spring. In agreement with Ref. [23], buds collected during spring and summer seasons produced a significantly higher percentage of explant establishment and were less contaminated than buds collected during autumn or winter.

### 3.2 Shoot propagation from the buds

As for surface sterilization of buds, different methods have been described. The disinfection procedures include washing in sodium hypochlorite (NaOCl) solution [24], mercuric chloride ($HgCl_2$) solution [19], 75% alcohol followed by $HgCl_2$ solution [25] or calcium hypochlorite [$(Ca(OCl)_2$] solution [26]. In our experiment, we used an $HgCl_2$ solution (1 $mg.l^{-1}$), which is highly efficient for surface sterilization of buds from field-grown trees.

As in other plant species, the optimal basal medium is often cultivar-dependent in *Malus* species [27]. Additionally, Kabylbekova et al. [28] have used response surface methodology (RSM) and showed that each apple cultivar requires a different composition of mineral nutrition for its optimal growth. Similarly, the selection of the optimal plant growth regulators is also genotype-dependent, as has been shown in the study of different apple scions and rootstocks [29, 30]. Thus, different media types for bud induction and shoot development were tested. The MS medium with 0.5 mg $l^{-1}$ BAP, 0.1 mg $l^{-1}$ IBA, 10 mg $l^{-1}$ glutamine and 2 mg $l^{-1}$ glycine was the most effective medium, based on monitoring following parameters: percentage of contamination, percentage of necrotic explants and percentage of explants with shoot initiation. In our experiment, plant growth regulators BAP and IBA were used for culture establishment and also for shoot multiplication, consistently with the studies from Refs. [31] or [28]. However, application of meta-topolin [32] or TDZ [33] in the growth medium was also described in apples.

### 3.3 Shoot proliferation

Newly developed shoots from bud explants were transferred into shoot induction media (**Figure 1**). Based on determination of multiplication parameters, that is, multiplication index (the number of newly formed shoots per initial shoot tip) and length of lateral shoots, MS medium consisted of 0.2 mg $l^{-1}$ BAP, 0.1 mg $l^{-1}$ IBA, 200 mg $l^{-1}$ glutamine, 2 mg $l^{-1}$ glycine and 200 mg $l^{-1}$ casein was chosen. Same as in our experiment, Sota et al. [34] have used MS medium for shoot multiplication of wild apples. However, the BAP concentration was quite higher than in our experiment (1 $mg.l^{-1}$ BAP), and instead of IBA, α-Naphthaleneacetic acid (NAA) was used. Additionally, the application of other growth nutrient media has also been reported [27, 35].

### 3.4 *In vitro* rooting, acclimatization and hardening

Well-multiplied shoots were subjected to MS medium lacking cytokinins but supplemented with auxins. All the treatments resulted in root production. The highest rooting percentage and roots per cultured shoot were obtained on ¼ MS supplemented with 0.5 mg $l^{-1}$ IBA, 10 g $l^{-1}$ sucrose and 6 g $l^{-1}$ agar. Although some authors found that NAA [36] or IAA [37] is more effective for rooting, the application of IBA showed the best results in our experiment.

The rooted shoots obtained from the best treatment (**Figure 2**) were removed from the rooting medium and the plantlets were then transferred to planter with perlite and watered with liquid MS medium. After 2 weeks, plantlets were transferred into bigger planter containing a mixture of perlite and soil and gradually adapted to lower moisture conditions. Finally, fully adapted plants (**Figure 3**) were transferred into greenhouse.

**Figure 1.**
*In vitro shoot multiplication of M. sylvestris L.*

### 3.5 Qualitative and quantitative traits of wild apples at experimental trials

The study was conducted at the Oldřichov (425 m a.s.l., central Bohemian Highlands) in 2003 and at Polná II (550 m a.s.l., Czech-Moravian Highlands) in 2007 in the Czech Republic. At both experimental sites, row planting with 2 × 2 m spacing was used and the plots were fenced off for the entire monitoring period. The subject of the evaluation of quantitative traits was survival rate, height and diameter at breast height (DBH). The DBH was measured using a millimetre calliper, and the height was determined using a measuring rod and a Vertex III ultrasonic altimeter (Haglöf Sweden AB, Langsele, Sweden). As for qualitative traits, trunk shape, forkness, branching angle, branch thickness and vitality were established. The qualitative traits were determined according to the manual in **Table 1**.

In Oldřichov, grafters of wild apples were planted together with *in vitro*-derived plantlets (**Figure 4**). For statistical evaluation, data from 2010 to 2018 were used

**Figure 2.**
*In vitro rooting of M. sylvestris L.*

**Figure 3.**
*Fully adapted in vitro-derived M. sylvestris L. plantlets.*

| **Trunk shape** | | **Forkness** | | **Branching angle** | | **Branch thickness** | | **Vitality** | |
|---|---|---|---|---|---|---|---|---|---|
| 1 | Straight | 1 | Stem continuous to the crown | 1 | Horizontal | 1 | Thin (up to 10% DBH) | 1 | Highly vital |
| 2 | Slightly curved on one side | 2 | Fork in the upper third of the tree's height | 2 | Ascending | 2 | Medium thick (10–25% DBH) | 2 | Vital |
| 3 | Strongly curved on one side | 3 | Fork in the second third of the tree's height | 3 | Overhanging | 3 | Thick (over 25% DBH) | 3 | Less vital |
| 4 | Trunk at least 2x Slightly ace-curved | 4 | Fork in the lower third of the tree | | | | | 4 | Declining tree |
| 5 | Significantly bent trunk | 5 | Repeated multiple forkness | | | | | | |

**Table 1.**
*The list of qualitative traits and their descriptors.*

**Figure 4.**
*M. sylvestris L. trees grown at Oldřichov research plot in 2010.*

(**Table 2**). The mortality of tissue culture plantlets was zero, while grafters' mortality reached 24%, mostly due to withering of the trees. ANOVA did not show any significant differences in height between *in vitro*-derived plantlets and grafters ($\alpha = 0.05$). However, in 2018, grafters were significantly higher than tissue culture plantlets, according to ANOVA, which could be caused by the growth rate of rootstock. No differences were observed between qualitative traits (**Table 2**). The trunk shape reached

| Group | 2010 | | | | 2018 | | | | | | | | | | | |
|---|---|---|---|---|---|---|---|---|---|---|---|---|---|---|---|---|
| | | Height (m) | | | | Height (m) | | | DBH (cm) | | | TrSh | Bra | BrAng | BrTh | Vit |
| | Count | Mean | Median | StD | Count | Mean | Median | StD | Mean | Median | StD | Median | Median | Median | Median | Median |
| *In vitro* | 15 | 3.03 | 3.25 | 0.90 | 15 | 6.89 | 7.35 | 2.95 | 5.24 | 5.80 | 1.61 | 3 | 3 | 2 | 2 | 1 |
| Grafters | 195 | 2.79 | 3.30 | 1.45 | 148 | 8.01 | 8.63 | 2.80 | 7.26 | 7.50 | 1.56 | 3 | 3 | 2 | 2 | 1 |
| All | 210 | 2.80 | 3.28 | 1.41 | 163 | 7.91 | 7.40 | 1.66 | 7.08 | 8.40 | 2.82 | 3 | 3 | 2 | 2 | 1 |

**Table 2.**
*Average growth parameters of Malus sylvestris L. plantlets at Oldřichov research plot. DBH - diameter of breast height, TrSh – trunk shape, Bra – Branching, BrAng – branching angle, BrTh – branch thickness, Vit - vitality and StD – standard deviation.*

| | 2008 | | | | 2017 | | | | | | | | | | | |
|---|---|---|---|---|---|---|---|---|---|---|---|---|---|---|---|---|
| | Height (m) | | | | Height (m) | | | | DBH (cm) | | | TrSh | Bra | BrAng | BrTh | Vit |
| Clone | Count | Mean | Median | StD | Count | Mean | Median | StD | Mean | Median | StD | Median | Median | Median | Median | Median |
| J1 | 9 | 1.14 $^{a, b}$ | 1.25 | 0.34 | 9 | 5.03 $^{a, b}$ | 4.9 | 1.00 | 4.16 | 4.05 | 0.93 | 3 | 4 | 2 | 3 | 1 |
| J2 | 10 | 1.21 $^{a, b}$ | 1.12 | 0.29 | 10 | 4.88 $^{a, b}$ | 4.95 | 0.68 | 3.96 | 3.93 | 1.01 | 3.5 | 3 | 2 | 3 | 1 |
| J4 | 23 | 0.98 $^{b}$ | 0.96 | 0.21 | 23 | 3.86 $^{b, c}$ | 3.7 | 0.78 | 3.14 | 3.05 | 1.14 | 3 | 3 | 2 | 3 | 1 |
| J5 | 18 | 1.11 $^{a, b}$ | 1.10 | 0.23 | 17 | 4.35 $^{a, b, c}$ | 4.6 | 0.72 | 3.99 | 4.1 | 1.11 | 3 | 3 | 2 | 3 | 1 |
| J6 | 50 | 1.15 $^{a, b}$ | 1.15 | 0.29 | 50 | 3.82 $^{c}$ | 3.75 | 1.07 | 3.48 | 3.35 | 1.71 | 3 | 3 | 2 | 3 | 1 |
| J7 | 18 | 0.99 $^{b}$ | 1.01 | 0.27 | 18 | 3.94 $^{b, c}$ | 4.05 | 0.61 | 3.26 | 3.4 | 1.23 | 3 | 3 | 2 | 3 | 1 |
| J8 | 12 | 1.13 $^{a, b}$ | 1.23 | 0.39 | 12 | 5.21 $^{a}$ | 4.9 | 0.61 | 4.43 | 4.28 | 0.61 | 3 | 3.5 | 2 | 3 | 1 |
| J9 | 10 | 1.20 $^{a, b}$ | 1.22 | 0.20 | 10 | 4.82 $^{a, b, c}$ | 4.8 | 1.03 | 3.83 | 4.1 | 1.39 | 3 | 3.5 | 2 | 3 | 1 |
| J10 | 14 | 0.89 $^{b}$ | 0.87 | 0.33 | 14 | 4.25 $^{a, b, c}$ | 4.15 | 1.09 | 3.84 | 4.28 | 1.70 | 3 | 3 | 2 | 3 | 1 |
| J12 | 20 | 1.10 $^{a, b}$ | 1.06 | 0.22 | 20 | 4.36 $^{a, b, c}$ | 4.5 | 0.88 | 3.27 | 3.3 | 1.37 | 3 | 3 | 2 | 2 | 1 |
| J13 | 19 | 1.10 $^{a, b}$ | 1.15 | 0.22 | 19 | 4.37 $^{a, b, c}$ | 4.4 | 0.77 | 3.53 | 3.55 | 1.18 | 3 | 3 | 2 | 3 | 1 |
| J26 | 20 | 1.33 $^{a}$ | 1.40 | 0.27 | 20 | 4.25 $^{a, b, c}$ | 4.4 | 1.06 | 3.68 | 3.45 | 1.62 | 3 | 4 | 2 | 3 | 1 |
| J28 | 13 | 1.14 $^{a, b}$ | 1.22 | 0.20 | 13 | 3.98 $^{b, c}$ | 3.8 | 0.78 | 4.51 | 4.7 | 1.63 | 3 | 3 | 2 | 3 | 1 |
| J30 | 14 | 1.12 $^{a, b}$ | 1.18 | 0.25 | 14 | 4.25 $^{a, b, c}$ | 4.4 | 0.94 | 4.27 | 4.45 | 1.33 | 3 | 3 | 2 | 3 | 1 |
| All | 250 | 1.11 | 1.12 | 0.28 | 249 | 4.24 | 4.30 | 0.96 | 3.69 | 3.75 | 1.42 | 3 | 3 | 2 | 3 | 1 |

**Table 3.**
*Average growth parameters of Malus sylvestris plantlets on Polná II research plot. DBH - diameter of breast height, TrSh - trunk shape, Bra - branching, BrAng - branching angle, BrTh - branch thickness, Vit - vitality and StD - standard deviation. Values followed by the same letter(s) (a-c) within the same column are not significantly different according to Tukey's test ($P < 0.05$).*

**Figure 5.**
*Malus sylvestris L. trees grown at Polná II research plot in 2011.*

the third degree (strongly curved on one side), same as forkness (forks in the second-third of the tree's height). The second degree was observed in the branching angle (ascending branches) and in the branch thickness (medium thickness with 10–25% DBH). The vitality reached the first degree (highly vital).

The growth characteristics of *in vitro*-derived wild apple plantlets recorded from years 2008 and 2017 grown at Polná II research plot are given in **Table 3**. The mortality was very low, only one individual in clone J5 died between the monitored years. ANOVA showed significant differences in height of wild apples in both years ($\alpha$ = 0.05), whereas no differences were observed in DBH. Among the qualitative traits, only small variances were noted. The trunk shape is on the third degree (strongly curved on one side) with the exception of clone J2, where the trunk was at least 2x slightly ace-curved. Forkness reached mostly the third degree. Only for clones J1 and J26, the fourth degree of forkness (forks in the lower third of the tree) was noted, and for clones J8 and J9, an intermediate stage between the third and the fourth degree was noted. All the wild apple clones had ascending branches (the second degree of branch angle). For clone J12, medium branch thickness was recorded (10–25% DBH), whereas all other clones reached the third degree of branch thickness (strong thickness with over 25% DBH). All tested clones were highly vital (**Figure 5**).

The comparison of our results with domestic or foreign studies is very problematic. There is a lack of evidence about the growth characteristics of wild apples, due to their low abundance in nature and difficult determination. Moreover, wild apples grown in our research plots are residual individuals selected in the Czech Republic, with different qualitative traits and growth potential.

## 4. Conclusions

The results of the presented study indicated that our methodology for the micropropagation of M. sylvestris L. from dormant bud could be efficiently used for *in vitro* conservation of endangered wild apples. Moreover, according to long-term experimental trials, *in vitro*-derived plantlets show quality and balanced growth, comparable with grafters. The growth characteristics of wild apples and grafters on the research plots at Oldřichov and Polná II will be still evaluated in the future.

In this regard, implementation of *in vitro* conservation of endangered wild apple M. sylvestris L. in practice can speed up the process of protection and reproduction.

## Acknowledgements

This research was supported by Institutional Support (MZE-RO0118).

## Conflict of interest

The authors declare no conflict of interest.

## Author details

Martina Komárková*, Helena Cvrčková, Jaroslav Dostál, Václav Buriánek and Pavlína Máchová
Forest and Game Management Research Institute, Jíloviště, Czech Republic

*Address all correspondence to: komarkova@vulhm.cz

## References

[1] Robinson JP, Harris SA, Juniper BE. Taxonomy of the genus Malus Mill. (Rosaceae) with emphasis on the cultivated apple, Malus domestica Borkh. Plant Systematics and Evolution. 2001;**226**:35-58. DOI: 10.1007/s006060170072

[2] Stephan R, Wagner I, Kleinschmit J. EUFORGEN Technical Guidelines for Genetic Conservation and Use for Wild Apple and Pear (Malus sylvestris and Pyrus pyraster). Italy: International Plant Genetic Recourses Institute; 2003

[3] Donno D, Beccaro GL, Mellano MG, Torello Marinoni D, Cerutti AK, Canterino S, et al. Application of sensory, nutraceutical and genetic techniques to create a quality profile of ancient apple cultivars. Journal of Food Quality. 2012;**35**:169-181

[4] Iacopini P, Camangi F, Stefani A, Sebastiani L. Antiradical potential of ancient Italian apple varieties of Malus×domestica Borkh. in a peroxynitrite-induced oxidative process. Journal of Food Composition and Analysis. 2010;**23**:518-524. DOI: 10.1016/j.jfca.2009.05.004

[5] Jakobek L, Barron AR. Ancient apple varieties from Croatia as a source of bioactive polyphenolic compounds. Journal of Food Composition and Analysis. 2016;**45**:9-15. DOI: 10.1016/j.jfca.2015.09.007

[6] Grulich V. Red List of vascular plants of the Czech Republic. Preslia. 2012;**84**:631-645

[7] Larsen AS, Asmussen CB, Coart E, Olrik DC, Kjær ED. Hybridization and genetic variation in Danish populations of European crab apple (Malus sylvestris). Tree Genetics & Genomes. 2006;**2**:86-97. DOI: 10.1007/s11295-005-0030-0

[8] Reim S, Proft A, Heinz S, Höfer M. Diversity of the European indigenous wild apple Malus sylvestris (L.) Mill. in the East Ore Mountains (Osterzgebirge), Germany: I. Morphological characterization. Genetic Resources and Crop Evolution. 2012;**59**:1101-1114. DOI: 10.1007/s10722-011-9746-x

[9] Wagner I, Maurer WD, Lemmen P, Schmitt HP, Wagner M, Binder M, et al. Hybridization and genetic diversity in wild apple (Malus sylvestris (L.) MILL.) from various regions in germany and from luxembourg. Silvae Genetica. 2014;**63**:81-93. DOI: 10.1515/sg-2014-0012

[10] Cornille A, Giraud T, Smulders MJM, Roldán-Ruiz I, Gladieux P. The domestication and evolutionary ecology of apples. Trends in Genetics. 2014;**30**:57-65. DOI: 10.1016/j.tig.2013.10.002

[11] Larsen AS, Jensen M, Kjær ED. Crossability between wild (Malus sylvestris) and cultivated (M. x domestica) apples. Silvae Genetica. 2008;**57**:127-130. DOI: 10.1515/sg-2008-0019

[12] Máchová P, Malá J, Cvrčková H, Dostál J, Buriánek V. In vitro reproduction of rare and endemic species of rowan tree. Journal of Forest Science. 2013;**2013**(59):386-390. DOI: 10.17221/46/2013-JFS

[13] Malá J, Máchová P, Cvrčková H, Karady M, Novak O, Mikulík J, et al. The role of cytokinins during micropropagation of wych elm. Biologia Plantarum. 2013;**57**:174-178. DOI: 10.1007/s10535-012-0252-6

[14] Pokorná E, Buriánek V, Máchová P, Dostál J, Komárková M. Nové poznatky o reprodukci topolu šedého v in vitro podmínkách. ZPRÁVY LESNICKÉHO VÝZKUMU. 2017;**62**:279-287

[15] Cvrčková H, Komárková M, Trčková O, Máchová P. Micropropagation of Tilia cordata Mill. and verification of genetic diversity of donor trees. In: Bonga JM, Park YS, Trontin JF, editors. In: Proceedings of the 5th International Conference of the IUFRO Unit 2.09.02 on "Clonal Trees in the Bioeconomy Age: Opportunities and Challenges." September 10-15, 2018. Coimbra, Portugal

[16] Machova P, Malá J, Cvrčková H. Micropropagation of Betula nana. Zprávy Lesnického Výzkumu. 2012;**57**:202-206

[17] Cvrčková H, Malá J, Machova P. In vitro reproduction of Jurinea cyanoides (L.) Rchb. Zprávy Lesnického Výzkumu. 2012;**57**:27-32

[18] Teixeira da Silva JA, Gulyás A, Magyar-Tábori K, Wang M-R, Wang Q-C, Dobránszki J. In vitro tissue culture of apple and other Malus species: Recent advances and applications. Planta. 2019;**249**:975-1006. DOI: 10.1007/s00425-019-03100-x

[19] Dobránszky J, Jámbor-Benczúr E, Hudák I, Magyar-Tábori K. Model experiments for establishment of in vitro culture by micrografting in apple. International Journal of Horticultural Science. 2005;**11**:47-49. DOI: 10.31421/IJHS/11/1/557

[20] Chokheli VA, Dmitriev PA, Rajput VD, Bakulin SD, Azarov AS, Varduni TV, et al. Recent development in micropropagation techniques for rare plant species. Plants. 2020;**9**:1733. DOI: 10.3390/plants9121733

[21] Murashige T, Skoog F. A revised medium for rapid growth and bio assays with tobacco tissue cultures. Physiologia Plantarum. 1962;**15**:473-497

[22] Kaplan Z, Danihelka J. Klíč ke květeně České republiky. Prague: Academia; 2019

[23] Modgil M, Sharma DR, Bhardwaj SV. Micropropagtion of apple cv. Tydeman's Early Worcester. Scientia Horticulturae. 1999;**81**:179-188. DOI: 10.1016/S0304-4238(98)00259-3

[24] Grant NJ, Hammatt N. Increased root and shoot production during micropropagation of cherry and apple rootstocks: effect of subculture frequency. Tree Physiology. 1999;**19**:899-903. DOI: 10.1093/treephys/19.13.899

[25] Wang Q, Tang H, Quan Y, Tang YQ. Phenol induced browning and establishment of shoot-tip expiants of'Fuji'apple and'Jinhua'pear cultured in vitro. Journal of Horticultural Science. 1994;**69**:833-839

[26] Yepes LM, Aldwinckle HS. Micropropagation of thirteen Malus cultivars and rootstocks, and effect of antibiotics on proliferation. Plant Growth Regulation. 1994;**15**:55-67

[27] Ciccotti A, Bisognin C, Battocletti I, Salvadori A, Herdemertens M, Jarausch W. Micropropagation of apple proliferation-resistant apomictic Malus sieboldii genotypes. Agronomy Research. 2008;**6**:445-458

[28] Kabylbekova B, Kovalchuk I, Mukhitdinova Z, Turdiyev T, Kairova G, Madiyeva G, et al. Reduced major minerals and increased minor nutrients improve micropropagation in three apple cultivars. Vitro Cell DevBiol-Plant. 2020;**56**:335-349. DOI: 10.1007/s11627-019-10019-1

[29] Dobránszky J, Abdul-Kader A, Magyar-Tábori K, Jámbor-Benczúr E, Bubán T, Szalai J, et al. In vitro shoot multiplication of apple: comparative response of three rootstocks to cytokinines and auxin. International Journal of Horticultural Science. 2000;**6**:36-39

[30] Dobránszky J, Abdul-Kader A, Magyar-Tábori K, Jámbor-Benczúr E, Bubán T, Szalai J, et al. Single and dual effects of different cytokinins on shoot multiplication of different apple scions. International Journal of Horticultural Science. 2000;**6**:76-78

[31] Boudabous M, Mars M, Marzougui N, Ferchichi A. Micropropagation of apple (Malus domestica L. cultivar Douce de Djerba) through in vitro culture of axillary buds. Acta Botanica Gallica. 2010;**157**:513-524

[32] Magyar-Tábori K, Dobránszki J, Jámbor-Benczúr E. High in vitro shoot proliferation in the apple cultivar Jonagold induced by benzyladenine analogues. Acta Agronomica Hungarica. 2002;**50**:191-195

[33] Bacha N, Darkazanli K, Abdul-Kader A. Direct organogenesis and plantlet multiplication from leaf explants of in vitro-grown shoots of apple (Malus domestica Borkh.) cv.'golden delicious' and 'MM111'rootstock. Fruit, Vegetable and Cereal Science and Biotechnology. 2009;**3**:28-34

[34] Sota V, Benelli C, Çuko B, Kongjika E. Effect of a double phase culture system and activated charcoal on in vitro propagation of Malus sylvestris (L.) Mill. Advances in Horticultural Science. 2022;**35**:361

[35] Muleo R, Morini S. Physiological dissection of blue and red light regulation of apical dominance and branching in M9 apple rootstock growing in vitro. Journal of Plant Physiology. 2008;**165**:1838-1846

[36] Zhang Y, Bozorov T, Li D, Zhou P, Wen X, Ding Y, et al. An efficient in vitro regeneration system from different wild apple (Malus sieversii) explants. Plant Methods. 2020;**16**:1-10

[37] Bommineni V, Mathews H, Samuel S, Kramer M, Wagner D. A new method for rapid in vitro propagation of apple and pear. HortScience. 2001;**36**:1102-1106. DOI: 10.21273/HORTSCI.36.6.1102

Section 2

# Hormonal Regulation, Phytopathogens and Phenolics

Chapter 4

# Deciphering the Plant Hormones Cross-Talk during Fruit Development: A Review

*Siti Khadijah A. Karim*

## Abstract

Horticultural industries are increasingly crucial in providing livelihoods, food quality, profits, and economic growth. In many horticultural plants, extensive studies were conducted to study the roles of hormones, epigenetics, and genes in regulating the development of cell number, cell size, fruit size, fruit weight, and endoreduplication primarily via a gene-mapping technique known as quantitative trait loci (QTL). In general, these plants encompassed those with full-genomes sequenced, such as the apple, tomato, strawberry, and bananas. However, apart from fully sequenced apple genomes, the genome sequences of many other plants, particularly highly profitable tropical fruits, such as mangoes, pineapples, durians, and coconuts are yet available. This chapter will describe the interplay of plant hormones in determining fruit cell number and cell size, which, in turn, affects the final fruit size in horticultural plants.

**Keywords:** fruit development, plant hormones, cross-talk, fruit size, hormonal control

## 1. Introduction

### 1.1 Origin of apple fruit cultivars

Apple (*Malus x domestica* Borkh.), also known as *M. pumila* [1, 2] belongs to the Rosaceae family [2–5]. Under the *Malus* genus, between 55 to 79 species have been identified. Each species can be further divided into different cultivars [6]. Based on the heterozygous nature of the genes as well as self-incompatibility, more than 1000 apple cultivars have been developed through crossing [2, 4, 7]. *M. domestica* is referred to as *Malus* spp. that had undergone the domestication process, while *M. sieversii* is referred to as the Central Asian 'wild apple', with the 'wild apple' term also referring to all apple species other than the domesticated apple. For instance, the most common 'wild apple' in Europe is *M. sylvestris* Miller, while *M. coronaria* (L.) Miller is a common species in North America [6].

## 2. Fruit development

### 2.1 Stages of fruit development

Fruiting bodies of the Angiosperms range from a completely dry fruit, such as *Arabidopsis*, a species of Brassicaceae, to a fleshy fruit such as the tomato (*Solanum lycopersicum*). Angiosperm plants include small plants up to large trees, and are the source of many major crops [8]. Dry fruits can either be indehiscent (split down one side to release their seeds) or non-dehiscent. Fleshy fruits can be classified into two categories: either true fruits (fruits derived from the carpel where the ovary wall expands into flesh) or false fruits (fruits derived from carpel and accessory structures) [8]. True fruits consist of a pericarp, which is obtained from the ovary wall, and seeds, which are derived from the fertilised ovules [9, 10]. *Arabidopsis* produces true fruits which develop from a gynoecium, consisting of two carpels that share a fused tissue called a septum [11], while tomato produces fleshy true fruits. Apple is a fleshy accessory fruit (false fruit). It is comprised of two distinct parts; an expanded ovary that develops as the "core" of the fruit, and a cortex, the edible part of the fruit, which develops from the fused base of the stamens, petals, and sepals called the hypanthium [1]. In *Arabidopsis*, growth of the embryo, the ovules and gynoecium involve cell differentiation and must be carefully coordinated during fruit development to generate the final product, being the fruit [11]. Most apple cultivars are not self-fruitful, meaning that they cannot produce a consistent yield without being fertilised by pollen from a different cv. [12]. After flowering and pollination, apple fruit development takes place over 20–21 weeks involving stages of fruit setting, cell division, cell expansion, maturation, and ripening which will lead to a crisp fruit with a waxy cuticle [10]. Apple flowers are produced in clusters consisting of five to six flowers; the primary flower, also known as "king flower", will normally open first and is capable of producing good quality fruit (**Figure 1**). However, the lateral flowers, which open later, can also produce good quality fruits. Apple trees are considered to reach full bloom after 50% of the king flowers on the tree are open [12].

Fruit development involves fruit set, cell division, cell expansion, and ripening. In accessory fruits, like apples, fruit set is the initiation phase of fruit development, encompassing development of the ovaries and its accessory organs into a rapidly growing young fruit. This happens during and soon after fertilisation has been

**Figure 1.**
*The positions of flowers in the cluster. King flower was positioned at the top of the cluster while at the lower positions were the lateral flowers.*

completed [13]. If fertilisation does not happen, the fruit will be abscised or be small, resulting in very few or no seeds [12]. A successful fertilisation process is influenced by several factors, including adequate pollination, nutrition, and optimum temperature [12]. Fruit produced without fertilisation could be an advantage for improvement of many crops as seedless fruits are easy to consume. Therefore, tomato mutants that exhibit parthenocarpic (seedless) fruit growth in the absence of fertilisation have been extensively studied [14].

For fruit set to take place, pollination and fertilisation must be accomplished or the flowers will be abscised [13, 15, 16]. Pollination is the crucial starting point in fruit development as it is required for fertilisation. During fruit set, rapid cell division occurs and then stops when the fruit reaches its mature size [16–18]. Levels of plant hormones, such as auxins, CKs (CKs), and gibberellic acids (GAs), increase to promote fruit set [13, 16]. In tomato, after successful fertilisation, ovary development into a fruit starts with cell division until roughly 10–14 days, and during the following six to seven weeks, the fruit grows by cell expansion [14]. During cell division, which starts during fruit setting and continues until three to four weeks after pollination, the concentrations of auxin, CK and GA increase to stimulate and support the process [14, 19, 20].

Cell division ceases to allow cell expansion to then take place. Auxin, gibberellin (GA), brassinosteroid (BR) and abscisic acid (ABA) levels rise during cell expansion [13], which continues until the end of ripening, reaches its peak at 40–60 days after anthesis and becomes constant during ripening [1]. Fruit ripening is a developmental process involving physiological and metabolic changes of colour, texture, aroma, and nutritional status [21]. These changes are helpful in promoting seed dispersal by making it attractive to birds and animals. With the onset of ripening, auxin stimulates ethylene production until reaching its optimum concentration, but it can also delay ripening [22, 23]. ABA and ethylene are produced during ripening to regulate ripening [22].

### 2.2 Cell differentiation

At the cellular level, fruit cells undergo a differentiation process in order to change into mature cells. The term differentiation is sometimes used in two different manners: 1) it may be used to describe the development of different specialised types of mature cells within an organ or tissue; 2) it can refer to the changes that occur during the development of a meristematic cell into a mature cell, which usually involves cell division and cell expansion [24]. For the purposes of this study, cell differentiation was defined as changes of undifferentiated meristematic cells into differentiated cells that divide, expand, and ripen. The early phase of apple fruit development involves rapid cell division where the cell number is greatly amplified prior to exit from cell division at approximately 24 days after full bloom (DAFB) [25]. Final fruit size is determined by the extent of cell division and cell expansion during growth and development. Therefore, understanding cell cycle regulation and cell expansion is important to the study of fruit development in plants [26].

Once cell proliferation has ceased, cells generally undergo an expansion process that results in water uptake into the central vacuole [24]. Plant cells are large, pressurised, and have a large central vacuole that allows accumulation of water and solutes, along with a strong cell wall [27]. The cell wall is made up of a mixture of carbohydrates and a small amount of proteins, making it rigid yet extensible [28]. During cell division, cells are divided from one cell into two cells while cell expansion occurs because of the expanded vacuole, ensuing in volume growth beyond the size of the mother cell before mitosis [29]. Water uptake and cell turgor are the main factors behind the expansion process. Though the cell size increased, it was generally assumed that little to no increase in cytoplasmic mass occur [30]. Rather, it is the vacuole that grows considerably in volume and the cell wall become thinner as it is stretched and new cell wall material is added to maintain wall thickness. To allow the cell wall to grow, transcripts encoding wall-loosening enzymes, such as polygalacturonase (PG), are up-regulated and newly deposited cell wall materials are hydrated resulting in relaxation and extensive thinning of the wall [31]. Other classes of cell wall-modifying enzymes/proteins are also involved in this process, such as pectin methylesterase (PME), β-galactosidase (β-gal), endo-1,4-β-glucanase (EGAe), xyloglucan endotransglucosylase/hydrolase (XTH), and expansins [32]. These enzymes also interact synergistically by promoting ripening through disassembling the pectin of the cell wall that results in fruit softening. The secondary wall is deposited after expansion has ceased, becoming impermeable and providing strength to cells and tissues [33].

## 3. Hormonal control during early fruit development

After fruit set, fruit undergo cell division and subsequently cell expansion stage. It is difficult to separate fruit set from subsequent stages of fruit development, though auxin and GA are indeed key in the sustained growth of fruit. Plant hormones such as CK and auxins are involved mostly in cell division and cell expansion during fruit development [34]. These hormones have been well-documented in stimulating cell cycle activity [13]. Auxin and GA co-regulate fruit set via auxin activation of GA synthesis [35]. Therefore, auxin and GA have been widely used to increase fruit set by inducing parthenocarpic growth in many crops [36]. For example, GA or auxin treatments to tomatoes can lead to parthenocarpic fruit development [17, 37–41]. Silencing of *SlARF7*, an auxin negative regulator, caused production of parthenocarpic fruits as a result of the increasing of auxin and GA concentrations, indicating an interaction between these hormones in regulating fruit set [35, 42]. Early studies showed that auxin concentrations increase during seed development and then GA concentrations increase in the ovaries during fruit set [13, 43, 44], and this is evidenced by the application of GA inhibitors to tomato which resulted in a decreased fruit set. High expression of GA-related genes in parthenocarpic fruit (*pat*) mutants (a recessive mutation conferring parthenocarpy in tomato) also supports that GA controls or influences tomato fruit set in parthenocarpic fruit [44].

In normal fruit development, successful pollination and fertilisation induces an increase in both auxin and GA concentrations within the ovary [14, 45–47]. Japanese pears produced larger fruits as a result of hand pollination, indicating that an increased number of fertilised stigma leads to higher levels of GA being produced by the pollen, therefore enhancing cell division and development of larger fruit [48]. This suggests that GA is critical in pollination and fertilisation [43]. Treatment of tomato ovaries with auxin causes the formation of fruits with a higher number of pericarp cells from cell division stimulation by auxin compared to GA-induced fruits that consist of fewer cell numbers but larger cells because of GA cell's enlargement role during fruit growth [13]. Normal-sized fruits undergo a balance cell division and cell expansion processes as a result of stable concentrations of both auxin and GA [49]. In addition, GA is also involved in the growth of certain fruits without the help of auxin and the outcome are fruit sizes that are twice as large, as shown in a transgenic tomato

(*S. lycopersicum* L.) with low levels of auxin response factor 7 (*Sl*ARF7) [17]. Commercially, GA is used to enhance fruit size and fruit cluster in parthenocarpic fruits by increasing carbohydrate import to the fruits, as parthenocarpic fruits are normally smaller and in compact fruit clusters, such as grapes, citrus, and berries [50]. GA has also been used to overcome fruit set problems of apple and pear trees particularly during biennial bearing (a phenomenon where the high production of fruits one year suppresses flower production of the coming year, hence lower yield production) to promote flower production thereby increasing fruit set and yield [50].

Another important function of GA in fruit development is to stimulate organ growth [51]. GA also features in germination, flowering, and fruit set in many plant species [40, 52, 53]. GA concentrations during fruit development increase twice; once during early fruit growth in order to trigger cell division and a second time during cell expansion [18]. During early development, GA is produced by pollen to facilitate pollen tube growth and germination [14]. As the pollen will transfer some GA into the ovary to trigger fruit growth [14, 38], an elevated GA concentration in the ovaries following pollination (which later cause auxin levels to increase [18]) suggests that both are involved together in fruit set and growth of tomato [40]. Studies on GA during cell division have been reported in hypocotyls of cucumber and tomato [48, 54]. GA is also reported to induce and maintain cell expansion [18, 48, 55] when auxin concentrations have decreased [18]. The function of GA in cell expansion is supported by the findings of larger cells in $GA_3$-induced fruit (parthenocarpic) compared to seeded fruit even though the fruit size was smaller than the seeded [14]. Application of 2,4-D and $GA_3$ together results in same size and shape of cells of parthenocarpic fruit as seeded fruit [14]. GA maintains cell expansion when it is used during the early stage of cell expansion in Japanese pears, resulting in larger fruits compared with untreated fruits [48, 56–58]. Meanwhile, in tracheid element differentiation, application of $GA_3$ in conjunction with auxin results in substantial tracheid expansion as GA causes the cell to expand while application of auxin only results in short tracheid growth [59]. This shows that both auxin and GA play a coordinated role in controlling cell division and cell expansion, probably based on a common response pathway. Early hypotheses speculated that GA may cause an increase in auxin biosynthesis or transport [60].

Meanwhile, the function of auxin during fruit set is demonstrated by its presence in pollen, its production in the stalk (style), and during fertilisation [14]. Its role in tomato fruit set regulation has been described by De Jong et al. [42] through the loss-of-function of *IAA9* and *ARF7*, resulting in parthenocarpic fruit growth. This suggests that auxin inhibits fruit growth until fertilisation takes place. *IAA9* is a tomato Aux/IAA transcriptional regulator that is linked to plant responses to auxin through the expression of auxin responsive genes [61]. The reduction of *IAA9* concentrations in tomato plants elicits pleiotropic phenotypes, indicating that *IAA9* acts as a transcriptional repressor of auxin signalling [42, 61]. In addition, another negative regulator of fruit set, *ARF8,* also had similar effects in *Arabidopsis ARF8* mutants [62], where *ARF8* caused suppression in ovary growth through a repressive action of the *Aux/IAA-ARF* complex on auxin responsive genes [63].

Besides fruit set, auxin is also known for promoting cell division and cell expansion [24]. While CKs control cyclin activity during the transition phase between $G_1$ to S during the cell cycle (which will be discussed further later), auxin's involvement in the cell cycle occurs much earlier than that by acting as a permissive signal for cell division to start [64]. However, this process is not yet fully understood [64–66]. Much research has been carried out focusing on identifying auxins` role in regulating cell differentiation in plants. For example, an auxin gene, auxin response factor (*ARF106*), is expressed during cell division and cell expansion with apple fruit development [19]. Since the gene was also co-localised with a fruit-size QTL, this suggests that auxin is involved in fruit growth control through cell differentiation [67]. Another example would be the auxin receptor, auxin binding protein (ABP1), which modulates ion fluxes in response to the hormone, and is proposed to mediate auxin-dependent cell expansion and is essential for cell division [19, 64]. Evidence has been made available in the form of applying antisense suppression of the *ABP1* gene in tobacco BY-2 cell cultures, resulting in slow proliferation, discarding auxin-induced cell expansion and reducing cell division [68]. Moreover, a mutation in the *ABP1* gene in *Arabidopsis* causes a lethal effect to cells [64]. Loss of function of ABP1 in *Arabidopsis* also resulted in lethal embryo as the result of cell expansion arrest, indicating auxin role, through the function of ABP1, in cell expansion stage during embryogenic development [68].

Along with auxin and GA, CK and ABA are also believed to be involved in fruit set. However, the cross-talk between them during fruit development is only partly understood. ABA is also involved in long-term developmental plant growth processes. While auxins, CKs, and BR are involved in early development, ABA is mainly present in the later stage of development where cell maturation transpires. During pollination of tomato fruit, where auxin and GA are present, ABA concentrations decrease while those of CK increase [69]. ABA concentrations continue to decrease shortly after pollination [69], and validated by a decrease in the mRNA levels of ABA biosynthesis genes after pollination [49] and the diminution of ABA concentrations in tomato pistils after pollination [70]. However, higher concentrations of ABA were observed in pollinated fruits compared to that in parthenocarpic fruits [71]. Despite being suppressed during and shortly after pollination in tomato fruit, ABA levels rose afterwards to support fruit set where they were detected about 5 days after pollination, enhanced in seed and pericarp until 30–50 days after pollination, with concentrations also increasing during cell expansion [20]. In rice (*Oryza sativa*) grains, low concentrations of ABA are seen in actively dividing endosperm cells [72], indicating its antagonistic effect in cell division. This is supported by higher ABA concentrations in less-dividing of small fruit Japanese pear 'Shinkou' compared to large Japanese pear cultivar 'Atago' during early fruit development [48]. In tomato, ABA shows a broad peak during cell expansion and cell maturation, showing ABA is involved in the cell expansion phase and reaches peak levels during the cell maturation phase [18]. ABA association in cell expansion has been determined by the reduction of fruit size in ABA-deficient mutants [73]. However, ABA's exact role in fruit development is not yet known.

Auxin role in regulating cell division by stimulating cell cycle progression has been discussed earlier. As for CKs, reducing CKs concentrations by overexpression of an inactivating enzyme [74, 75] and insensitivity to the hormone cause dwarfism because of reduced cell numbers in *Arabidopsis* [76]. CK role in cell expansion was reported in *Arabidopsis* leaf expansion which resulted from cell expansion [77]. A point to note, CK function in cell division and cell expansion might require auxin. For example, in *Zinnia elegans* cell cultures, combination of auxin and CK was required to induce cell division and cell differentiation [33, 78]. Even though CKs are generally considered vital in the stimulation of cell division during fruit development [71], very little experimental data supports their involvement in the initial cell division phase of fruit growth [20]. However, it is known that cyclin D activity in plants is influenced by hormones and carbohydrate levels [79]. In tobacco, it has been shown that overexpression of the cell cycle stimulator *CyclinD2* accelerated plant growth [80]. Riou-Khamlichi et al. [81] concluded that cell division can be induced and maintained in the absence of exogenous CK in transgenic plants over-expressing *CycD3* (CDK4 to CDK6). Specifically, with overexpression of *CycD3*, calli may be cultured without the presence of CK [24]. This is further underscored by the transcription of *CycD3* in *Arabidopsis* suspension culture after it is added with CK and sucrose while expression of cyclin D2 and D4 were induced in the presence of sucrose only [79]. Overexpressing *CycD3* also reduces endoreduplication [82, 83] while loss of *CycD3* function induces endoreduplication in *Arabidopsis* [26, 84].

BRs play an important role in early fruit development by promoting cell division, cell expansion in the stem, inhibiting root growth, promote xylem differentiation, ripening, and abscission [85–88] such as in tomato [89, 90], grape berry [91], and cucumber [36, 92]. In tomato fruit, treatment with BR restores the dry mass content, sugar, and amino acid levels in dwarf tomato mutants, showing that BR is required for tomato fruit development [89, 92]. In cucumber, the application of exogenous BR to a cultivar without parthenocarpic capacity induces parthenocarpic growth and increases cell division via cell cycle-related gene expression [36]. In contrast, application of a BR biosynthesis inhibitor (brassinazole (Brz)) to a cucumber cultivar with parthenocarpic growth blocks fruit set, whereas this inhibitory effect was subsequently reversed by applying exogenous BR [36].

## 4. Fruit ripening, ethylene biosynthesis, and other related hormones

In general, there are two types of fruits based on the rate of respiration. Non-climacteric fruits do not increase ethylene production when they ripen [93]. Climacteric fruits require synthesis, perception, and signal transduction of the plant hormone ethylene in order to fully ripen [94]. In climacteric fruits, when ethylene is detected, they undergo autocatalytic ethylene biosynthesis and increased cell respiration, denoted by increased $CO_2$ production, which leads to fruit ripening [93]. Non-climacteric fruit, such as strawberries, grapes and citrus, require ethylene for certain ripening processes, such as skin de-greening, and do not undergo the autocatalytic response observed in climacteric fruit [94].

The inability of transgenic tomato plants to ripen where ethylene biosynthesis has been disrupted shows that the autocatalytic burst of ethylene biosynthesis is essential for climacteric fruit ripening [95]. Nevertheless, the distinction between climacteric and non-climacteric is not clearly delimited; closely related species of melon can be either climacteric, such as the cantaloupes, or non-climacteric like the 'honey dew' melon [94]. Climacteric fruit spans an evolutionarily wide range of angiosperm taxa, from eudicotyledons, such as tomato and apple, to monocotyledons like the banana [96]. Climacteric fruits are advantageous from an economic perspective as they are able to continue ripening even when removed from the plant [97]. This allows the fruit to be harvested and transported before they are fully ripe, resulting in reduced loss through fruit spoilage in transit [97]. Many non-climacteric fruit lack this ability and must ripen on the plant [97].

In the ethylene biosynthetic pathway, L-methionine is the main precursor which is then transformed into ethylene by the enzymes 1-aminocyclopropane-1-carboxylate (ACC) synthase (ACS) and ACC oxidase (ACO). ACS and ACO are encoded by multigene families in various plant species [98]. In addition, ethylene production can be brought about by using propylene with climacteric fruit (but not in non-climacteric type fruit). In apple fruit, the level of *ACS* and *ACO* transcripts increase at the onset of ripening after treatment with ethylene [99]. However, these genes are not expressed in non-climacteric fruits as these fruits produce ethylene in very low concentrations [100]. Nevertheless, ethylene production is induced in non-climacteric fruits by various external stimuli, such as physical wounding, auxin treatments, chilling injury, drought, water logging, and pathogen infections [101]. These observations show that non-climacteric fruits have the capability to produce ethylene but are not able to produce ripening-associated ethylene [101–103].

A number of investigations have been carried out on the ripening process at both the biochemical and genetic levels in many fruits, with tomato being the classical model of choice for fleshy fruit ripening [94, 96]. As tomato ripens, it undergoes a colour change from green to red through the transformation of chloroplasts to chromoplasts [104]. Then, the fruit softens and its texture changes as the fruit cell wall is modified and partially disassembled by enzymes [104]. Alteration of specific volatiles concentrations and the sugar-acid balance will lead to the development of flavours associated with a ripened tomato [104]. Similar changes occur during ethylene-induced ripening in apples and a number of other fruits [104]. When fruits ripen, cellular turgor pressure is decreased, the cell wall is dissembled, and cell adhesion is reduced, resulting in fruit softening; a process which is facilitated by many classes of cell wall-modifying enzymes such as PG, PME, β-gal, XTH, and expansin [32, 99].

ABA concentrations are very low in unripe fruit but increase when the fruit ripens [105]. This happens in both climacteric fruits, such as mango [93], pear, avocado, and apple [106], and non-climacteric fruits like citrus [107], cherry [108], and grapes [109]. In climacteric fruits like apples, ABA concentrations increase from maturation to harvest, while in non-climacteric sweet cherries, ABA concentrations increase before maturation and decrease until harvest [105, 110]. ABA appears to have a similar function to ethylene in regulating the changes of fruit during ripening by supporting the colour changes, cell wall metabolism, fruit softening, and sugar and acid metabolism [105].

In climacteric tomato fruit, BR has been reported to promote the ripening process through increasing ethylene production and lowering chlorophyll levels [90, 92, 93]. However, in mango, low concentrations of castasterone and brassinolide are present throughout the ripening period indicating that BR is unlikely to modulate ripening [93]. Chai et al. [92] demonstrated that BRs are involved in strawberry fruit ripening through downregulating the BR receptor gene (*FaBRI1*) transcript, producing plants that lack the red colouring traditionally associated with 'Akihime' strawberry fruit. As a result of these inconsistent findings, definitive evidence for action of this hormone in fruit ripening is still inadequate.

## 5. Conclusions

The molecular mechanisms controlling fruit growth and fruit size in apples have yet to be fully understood [26]. The recent publication of the genome sequence of apple provides a powerful tool to reveal the underlying mechanisms during fruit development [5].

## Acknowledgements

I would like to express appreciation to the University of Auckland, New Zealand, for my doctorate studies opportunity. An appreciation also goes to Malaysia Ministry of Education and Universiti Teknologi MARA (UiTM) for my doctorate financial support.

## Conflict of interest

I declare no conflict of interest.

## Author details

Siti Khadijah A. Karim
Faculty of Plantation and Agrotechnology, Universiti Teknologi MARA Melaka, Melaka, Malaysia

*Address all correspondence to: khadijahkarim@uitm.edu.my

## References

[1] Janssen BJ, Thodey K, Schaffer RJ, Alba R, Balakrishnan L, Bishop R, et al. Global gene expression analysis of apple fruit development from the floral bud to ripe fruit. BMC Plant Biology. 2008a;**8** (1):1-29

[2] Liu Z, Ma H, Jung S, Main D, Guo L. Developmental mechanisms of fleshy fruit diversity in Rosaceae. Annual Review of Plant Biology. 2020;**71**:547-573

[3] Faize M, Sourice S, Dupuis F, Parisi L, Gautier MF, Chevreau E. Expression of wheat puroindoline-b reduces scab susceptibility in transgenic apple (Malus×domestica Borkh.). Plant Science. 2004;**167**:347-354. DOI: 10.1016/j.plantsci.2004.04.003

[4] Khan SA, Tikunov Y, Chibon P-Y, Maliepaard C, Beekwilder J, Jacobsen E, et al. Metabolic diversity in apple germplasm. Plant Breeding. 2014;**133**: 281-290. DOI: 10.1111/pbr.12134

[5] Velasco R, Zharkikh A, Affourtit J, Dhingra A, Cestaro A, Kalyanaraman A, et al. The genome of the domesticated apple (Malus × domestica Borkh.). Nature Genetics. 2010;**42**:833-839

[6] Harris SA, Robinson JP, Juniper BE. Genetic clues to the origin of the apple. Trends in Genetics. 2002;**18**:426-430. DOI: 10.1016/S0168-9525(02)02689-6

[7] Chen X, Feng T, Zhang Y, He T, Feng J, Zhang C. Genetic diversity of volatile components in Xinjiang wild apple (Malus sieversii). Journal of Genetics and Genomics. 2007;**34**:171-179

[8] Seymour G, Poole M, Manning K, King GJ. Genetics and epigenetics of fruit development and ripening. Current Opinion in Plant Biology. 2008; **11**:58-63

[9] Cheniclet C, Rong WY, Causse M, Frangne N, Bolling L, Bordeaux VS, et al. Cell expansion and endoreduplication show a large genetic variability in pericarp and contribute strongly to tomato fruit growth. Plant Physiology. 2005;**139**:1984-1994. DOI: 10.1104/pp.105.068767.1984

[10] Janssen BJ, Thodey K, Schaffer RJ, Alba R, Balakrishnan L, Bishop R, et al. Global gene expression analysis of apple fruit development from the floral bud to ripe fruit. BMC Plant Biology. 2008b;**8**: 16. DOI: 10.1186/1471-2229-8-16

[11] Gu Q, Ferrándiz C, Yanofsky MF, Martienssen R. The FRUITFULL MADS-box gene mediates cell differentiation during Arabidopsis fruit development. Development. 1998;**125**:1509-1517

[12] Bekey RS. Pollination and Fruit Set of Apple. Corvallis, Or: Extension Service, Oregon State University; 1981

[13] McAtee P, Karim S, Schaffer R, David K. A dynamic interplay between hormones is required for fruit development, maturation, and ripening. Frontiers in Plant Science. 2013;**4**:79

[14] De Jong M, Mariani C, Vriezen WH. The role of auxin and gibberellin in tomato fruit set. Journal of Experimental Botany. 2009a;**60**:1523-1532. DOI: 10.1093/jxb/erp094

[15] Vivian-Smith A, Luo M, Chaudhury A, Koltunow A. Fruit development is actively restricted in the absence of fertilization in Arabidopsis. Development. 2001;**128**:3

[16] Karim SKA, Allan AC, Schaffer RJ, David KM. Cell division controls final fruit size in three apple (*Malus* x

*domestica*) cultivars. Horticulturae. 2022; **8**(7):657

[17] De Jong M, Wolters-Arts M, García-Martínez JL, Mariani C, Vriezen WH. The Solanum lycopersicum AUXIN RESPONSE FACTOR 7 (SlARF7) mediates cross-talk between auxin and gibberellin signalling during tomato fruit set and development. Journal of Experimental Botany. 2011;**62**:617-626. DOI: 10.1093/jxb/erq293

[18] Gillaspy G, Ben-david H, Gruissem W, Darwin C. Fruits: A developmental perspective. The Plant Cell. 1993;**5**:1439-1451

[19] Devoghalaere F, Doucen T, Guitton B, Keeling J, Payne W, Ling TJ, et al. A genomics approach to understanding the role of auxin in apple (Malus x domestica) fruit size control. BMC Plant Biology. 2012;**12**:7

[20] Mariotti L, Picciarelli P, Lombardi L, Ceccarelli N. Fruit-set and early fruit growth in tomato are associated with increases in indoleacetic acid, CK, and bioactive gibberellin contents. Journal of Plant Growth Regulation. 2011;**30**:405-415

[21] Schaffer RJ, Friel EN, Souleyre EJF, Bolitho K, Thodey K, Ledger S, et al. A genomics approach reveals that aroma production in apple is controlled by ethylene predominantly at the final step in each biosynthetic pathway. Plant Physiology. 2007;**144**:1899-1912

[22] Giovannoni JJ. Fruit ripening mutants yield insights into ripening control. Current Opinion in Plant Biology. 2007;**10**:283-289. DOI: 10.1016/j.pbi.2007.04.008

[23] Karlova R, Chapman N, David K, Angenent GC, Seymour GB, de Maagd RA. Transcriptional control of fleshy fruit development and ripening. Journal of Experimental Botany. 2014;**65**:4527-4541. DOI: 10.1093/jxb/eru316

[24] Anastasiou E, Lenhard M. Control of plant organ size. In: Bogre L, Beemster G, editors. Plant Growth Signaling. Plant Cell Monographs. Berlin: Springer-Verlag; 2008. pp. 25-45

[25] Malladi A, Johnson LK. Expression profiling of cell cycle genes reveals key facilitators of cell production during carpel development, fruit set, and fruit growth in apple (Malus x domestica Borkh.). Journal of Experimental Botany. 2011;**62**:205-219. DOI: 10.1093/jxb/erq258

[26] Malladi A, Hirst PM. Increase in fruit size of a spontaneous mutant of "gala" apple (Malus x domestica Borkh.) is facilitated by altered cell production and enhanced cell size. Journal of Experimental Botany. 2010;**61**:3003-3013. DOI: 10.1093/jxb/erq134

[27] Cosgrove DJ. Loosening of plant cell walls by expansins. Nature. 2000b;**407**:321-326

[28] Li Y, Jones L, McQueen-Mason S. Expansins and cell growth. Current Opinion in Plant Biology. 2003b;**6**:603-610. DOI: 10.1016/j.pbi.2003.09.003

[29] Verbelen J-P, Vissenberg K. Cell expansion: Past, present and perspectives. In: Robinson DG, Nick P, editors. Plant Cell Monographs: The Expanding Cell. Berlin Heidelberg: Springer; 2007. pp. 1-6. DOI: 10.1007/7089_2006_068

[30] Sablowski R, Dornelas MC. Interplay between cell growth and cell cycle in plants. Journal of Experimental Botany. 2014;**65**:2703-2714

[31] Wolf S, Hématy K, Höfte H. Growth control and cell wall signaling in plants.

Annual Review of Plant Biology. 2012; **63**:381-407

[32] Rugkong A, McQuinn R, Giovannoni JJ, Rose JKC, Watkins CB. Expression of ripening-related genes in cold-stored tomato fruit. Postharvest Biology and Technology. 2011;**61**:1-14. DOI: 10.1016/j.postharvbio.2011.02.009

[33] Church DL. Tracheid element element differentiation in Zinnia mesophyll cell cultures. Plant Growth Regulation. 1993;**12**:179-188. DOI: 10.1007/BF00027197

[34] Swarup R, Parry G, Graham N, Allen T, Bennett M. Auxin cross-talk: Integration of signalling pathways to control plant development. Plant Molecular Biology. 2002;**49**:409-424

[35] Kumar R, Khurana A, Sharma AK. Role of plant hormones and their interplay in development and ripening of fleshy fruits. Journal of Experimental Botany. 2014;**65**:4561-4575. DOI: 10.1093/jxb/eru277

[36] Fu FQ, Mao WH, Shi K, Zhou YH, Asami T, Yu JQ. A role of brassinosteroids in early fruit development in cucumber. Journal of Experimental Botany. 2008;**59**:2299-2308

[37] Bünger-Kibler S, Bangerth F. Relationship between cell number, cell size and fruit size of seeded fruits of tomato (Lycopersicon esculentum Mill.), and those induced parthenocarpically by the application of plant growth regulators. Plant Growth Regulation. 1982;**1**:143-154

[38] Gustafson FG. Parthenocarpy induced by pollen extracts. American Journal of Botany. 1937;**24**:102-107

[39] Gustafson FG. Influence of gibberellic acid on setting and development of fruits in tomato. Plant Physiology. 1960;**35**:521-523

[40] Serrani JC, Fos M, Atarés A, García-Martínez JL. Effect of gibberellin and auxin on parthenocarpic fruit growth induction in the cv. Micro-tom of tomato. Journal of Plant Growth Regulator. 2007;**26**:211-221

[41] Wittwer SH, Bukovac MJ, Sell HM, Weller LE. Some effects of gibberellin on flowering and fruit setting. Plant Physiology. 1957;**32**:39-41

[42] De Jong M, Wolters-Arts M, Feron R, Mariani C, Vriezen WH. The Solanum lycopersicum auxin response factor 7 (SlARF7) regulates auxin signaling during tomato fruit set and development. The Plant Journal. 2009b; **57**:160-170

[43] Jianhon H, Mitchum MG, Barnaby N, Ayele BT, Ogawa M, Nam E, et al. Potential sites of bioactive gibberellin production during reproductive growth in Arabidopsis. The Plant Cell. 2008;**20**: 320-336

[44] Olimpieri I, Siligato F, Caccia R, Soressi G-P, Mazzucato A, Mariotti L, et al. Tomato fruit set driven by pollination or by the parthenocarpic fruit allele are mediated by transcriptionally regulated gibberellin biosynthesis. Planta. 2007; **226**:877-888

[45] Koshioka M, Nishijima T, Yamazaki H, Liu Y, Nonaka M, Mander LN. Analysis of gibberellins in growing fruits of Lycopersicon Esculentum after pollination or treatment with 4-Chlorophenoxyacetic acid. Journal of Horticulture Science. 1994;**69**:171-179

[46] Mapelli S, Frova C, Torti G, Soressi GP. Relationship between set, development and activities of growth

regulators in tomato fruits. Plant and Cell Physiology. 1978;**19**:1281-1288

[47] Sjut V, Bangerth F. Induced parthenocarpy—A way of changing the levels of endogenous hormones in tomato fruits (Lycopersicon esculentum M.). Extractable hormones. Plant Growth Regulation. 1982;**1**:243-251

[48] Zhang C. Biologically active gibberellins and abscisic acid in fruit of two late-maturing Japanese pear cultivars with contrasting fruit size. Journal of American Society for Horticultural Science. 2007;**132**:4

[49] Vriezen WH, Feron R, Maretto F, Keijman J, Mariani C. Changes in tomato ovary transcriptome demonstrate complex hormonal regulation of fruit set. New Phytologist. 2008;**177**:60-76

[50] Taiz L. In: Taiz L, Zeiger E, editors. Plant Physiology. Sunderland, U.K: Mass Sinauer Associates; 2010

[51] Hedden P, Thomas SG. Gibberellin biosynthesis and its regulation. Biochemical Journal. 2012;**444**:11-25

[52] Hedden P, Kamiya Y. Gibberellin biosynthesis: Enzymes, genes and their regulation. Annual Review of Plant Physiology and Plant Molecular Biology. 1997;**48**:431-460

[53] Olszewski N, Sun T-P, Gubler F. Gibberellin signaling: Biosynthesis, catabolism and response pathways. The Plant Cell. 2002;**14**:61-80

[54] Asahina M, Iwai H, Kikuchi A, Yamaguchi S, Kamiya Y, Kamada H, et al. Gibberellin produced in the cotyledon is required for cell division during tissue reunion in the cortex of cut cucumber and tomato hypocotyls. Plant Physiology. 2002;**129**:201-210

[55] Ozga JA, Reinecke DM. Hormonal interactions in fruit development. Journal of Plant Growth Regulation. 2003;**22**:73-81

[56] Hayashi S, Tanabe K. Basic Knowledge of Fruit Tree Culture. Tottori, Japan: Association Agriculture Press; 1991

[57] Zhang C, Tanabe K, Tamura F, Itai A, Wang S. Spur characteristics, fruit growth, and carbon partitioning in two late-maturing Japanese pear (Pyrus pyrifolia Nakai) cultivars with contrasting. Journal of American Society for Horticultural Science. 2005;**130**:252-260

[58] Zhang C, Tanabe K, Tamura F, Itai A, Yoshida M. Roles of gibberellins in increasing sink demand in Japanese pear fruit during rapid fruit growth. Plant Growth Regulation. 2007;**52**:161-172

[59] Kalev N, Aloni R. Role of auxin and gibberellin in regenerative differentiation of tracheids in Pinus pinea seedlings. New Phytology. 1998; **138**:461-468

[60] Law DM, Hamilton RH. Effects of gibberellic acid on endogenous indole-3-acetic acid and indoleacetyl aspartic acid levels in a dwarf pea. Plant Physiology. 1984;**75**:255-256

[61] Wang H, Jones B, Li Z, Frasse P, Delalande C, Regad F, et al. The tomato Aux/IAA transcription factor IAA9 is involved in fruit development and leaf morphogenesis. The Plant Cell Online. 2005;**17**:2676-2692

[62] Goetz M, Vivian-smith A, Johnson SD, Koltunow AM. AUXIN RESPONSE FACTOR8 is a negative regulator of fruit initiation in Arabidopsis. The Plant Cell. 2006;**18**:5. DOI: 10.1105/tpc.105.037192.2

[63] Pandolfini T, Molesini B, Spena A. Molecular dissection of the role of auxin in fruit initiation. Trends in Plant Science. 2007;**12**:327-329

[64] David KM, Couch D, Braun N, Brown S, Grosclaude J, Perrot-Rechenmann C. The auxin-binding protein 1 is essential for the control of cell cycle. The Plant Journal. 2007;**50**: 197-206

[65] den Boer BG, Murray JA. Triggering the cell cycle in plants. Trends in Cell Biology. 2000;**10**:245-250

[66] Stals H, Inzé D. When plant cells decide to divide. Trends in Plant Science. 2001;**6**:359-364

[67] Dash M, Malladi A. The AINTEGUMENTA genes, MdANT1 and MdANT2, are associated with the regulation of cell production during fruit growth in apple (Malus × domestica Borkh.). BMC Plant Biology. 2012;**12**:98. DOI: 10.1186/1471-2229-12-98

[68] Chen JG, Ullah H, Young JC, Sussman MR, Jones AM. ABP1 is required for organized cell elongation and division in Arabidopsis embryogenesis. Genes & Development. 2001;**15**:6. DOI: 10.1101/gad.866201

[69] Kojima K, Sakurai N, Tsurusaki K. IAA distribution within tomato flower and fruit. HortScience. 1994;**29**: 1200

[70] Kojima K, Kuraishi S, Sakurai N, Fusao K. Distribution of abscisic acid in different parts of the reproductive organs of tomato. Scientia Horticulturae. 1993;**56**:23-30

[71] Srivastava A, Handa AK. Hormonal regulation of tomato fruit development: A molecular perspective. Journal of Plant Growth Regulation. 2005;**24**:67-82

[72] Yang J, Zhang J, Wang Z, Zhu Q. Hormones in the grains in relation to sink strength and postanthesis development of spikelets in rice. Plant Growth Regulation. 2003;**41**:185-195

[73] Nitsch L, Kohlen W, Oplaat C, Charnikhova T, Cristescu S, Michieli P, et al. ABA-deficiency results in reduced plant and fruit size in tomato. Journal of Plant Physiology. 2012;**169**:878-883

[74] Werner T, Motyka V, Laucou V, Smets R, Van Onckelen H, Schmülling T. CK-deficient transgenic Arabidopsis plants show multiple developmental alterations indicating opposite functions of CKs in the regulation of shoot and root meristem activity. The Plant Cell Online. 2003;**15**:2532-2550

[75] Werner T, Motyka V, Strnad M, Schmülling T. Regulation of plant growth by CK. Proceedings of the National Academy of Sciences. 2001;**98**: 10487-10492

[76] Riefler M, Novak O, Strnad M, Schmülling T. Arabidopsis CK receptor mutants reveal functions in shoot growth, leaf senescence, seed size, germination, root development and CK metabolism. The Plant Cell Online. 2006;**18**:40-54

[77] Letham D. Regulators of cell division in plant tissues. Physiologia Plantarum. 1974;**32**:66-70

[78] Fukuda H. Signals that control plant vascular cell differentiation. Nature Reviews: Molecular Cell Biology. 2004;**5**: 379-391. DOI: 10.1038/nrm1364

[79] Gutiérrez R, Quiroz-Figueroa F, Vázquez-Ramos JM. Maize cyclin D2 expression, associated kinase activity and effect of hormones during germination. Plant & Cell Physiology.

2005;**46**:166-173. DOI: 10.1093/pcp/pci007

[80] Cockcroft CE, den Boer BGW, Healy JMS, Murray JAH. Cyclin D control of growth rate in plants. Nature. 2000;**405**: 575-579

[81] Riou-Khamlichi C, Huntley R, Jacqmard A, Murray JAH. CK activation of Arabidopsis cell division through a D-type cyclin. Science. 1999;**283**:1541-1544

[82] Dewitte W, Riou-khamlichi C, Scofield S, Healy JMS, Jacqmard A, Kilby NJ, et al. Altered cell cycle distribution, hyperplasia, and inhibited differentiation in Arabidopsis caused by the D-type cyclin CYCD3. The Plant Cell. 2003;**15**:7. DOI: 10.1105/tpc.004838.et

[83] Schnittger A, Schöbinger U, Bouyer D, Weinl C, Stierhof Y-D, Hülskamp M. Ectopic D-type cyclin expression induces not only DNA replication but also cell division in Arabidopsis trichomes. Proceedings of the National Academy of Sciences. 2002;**99**:6410-6415

[84] Dewitte W, Scofield S, Alcasabas AA, Maughan SC, Menges M, Braun N, et al. Arabidopsis CYCD3 D-type cyclins link cell proliferation and endocycles and are rate-limiting for CK responses. Proceedings of the National Academy of Sciences. 2007;**104**:14537-14542

[85] Clouse SD. Arabidopsis mutants reveal multiple roles for sterols in plant development. The Plant Cell Online. 2002;**14**:1995-2000

[86] Hasan SA, Hayat S, Ahmad A. Brassinosteroids protect photosynthetic machinery against the cadmium induced oxidative stress in two tomato cultivars. Chemosphere. 2011;**84**:1446-1451

[87] Mandava NB. Plant growth-promoting brassinosteroids. Annual Review of Plant Physiology and Plant Molecular Biology. 1988;**39**:23-52

[88] Nemhauser JL, Mockler TC, Chory J. Interdependency of brassinosteroid and auxin signaling in Arabidopsis. PLoS Biology. 2004;**2**:258

[89] Lisso J, Altmann T, Müssig C. Metabolic changes in fruits of the tomato dx mutant. Phytochemistry. 2006;**67**: 2232-2238

[90] Vidya VB, Rao SSR. Acceleration of ripening of tomato pericarp discs by brassinosteroids. Phytochemistry. 2002; **61**:843-847

[91] Symons GM, Davies C, Shavrukov Y, Dry IB, Reid JB, Thomas MR. Grapes on steroids: Brassinosteroids are involved in grape berry ripening. Plant Physiology. 2006;**140**:150-158

[92] Chai, Y-M., Zhang, Q., Tian, L., Li, C-L., Xing, Y., Qin, L., & Shen, Y-Y. (2013). Brassinosteroid is involved in strawberry fruit ripening. Plant Growth Regulation 69, 63-69

[93] Zaharah SS, Singh Z, Symons GM, Reid JB. Role of brassinosteroids, ethylene, abscisic acid and indole-3-acetic acid in mango fruit ripening. Journal of Plant Growth Regulation. 2012;**31**:363-372

[94] Barry CS, Giovannoni JJ. Ethylene and fruit ripening. Journal of Plant Growth Regulation. 2007;**26**:143-159

[95] Min-Wong L, Oeller PW, Pike DA, Taylor LP, Theologis A. Reversible inhibition of tomato fruit senescence by antisense RNA. Science. 1991;**254**:437

[96] Giovannoni JJ. Genetic regulation of fruit development and ripening. The Plant Cell Online. 2004;**16**:S170-S180

[97] Watkins CB. Ethylene synthesis, mode of action, consequences and control. In: Knee M, editor. Fruit Quality and Its Biological Basis. CRC Press; 2002. p. 180

[98] Atkinson RG, Gunaseelan K, Wang MY, Luo L, Wang T, Norling CL, et al. Dissecting the role of climacteric ethylene in kiwifruit (Actinidia chinensis) ripening using a 1-aminocyclopropane-1-carboxylic acid oxidase knockdown line. Journal of Experimental Botany. 2011;**62**:3821-3835. DOI: 10.1093/jxb/err063

[99] Ireland HS, Gunaseelan K, Muddumage R, Tacken EJ, Putterill J, Johnston JW, et al. Ethylene regulates apple (Malus x domestica) fruit softening through a dose × time-dependent mechanism and through differential sensitivities and dependencies of cell wall-modifying genes. Plant Cell Physiology. 2014;**55**: 1005-1016

[100] Trainotti L, Pavanello A, Casadoro G. Different ethylene receptors show an increased expression during the ripening of strawberries: Does such an increment imply a role for ethylene in the ripening of these non-climacteric fruits? Journal of Experimental Botany. 2005;**56**:2037-2046. DOI: 10.1093/jxb/eri202

[101] Yamane M, Abe D, Yasui S, Yokotani N, Kimata W, Ushijima K, et al. Differential expression of ethylene biosynthetic genes in climacteric and non-climacteric Chinese pear fruit. Postharvest Biology and Technology. 2007;**44**:220-227

[102] Theologis A. One rotten apple spoils the whole bushel: The role of ethylene in fruit ripening. Cell. 1992;**70**:181-184

[103] Yang SF, Hoffman NE. Ethylene biosynthesis and its regulation in higher plants. Annual Review of Plant Physiology. 1984;**35**:155-189

[104] Alexander L, Grierson D. Ethylene biosynthesis and action in tomato: A model for climacteric fruit ripening. Journal of Experimental Botany. 2002; **53**:2039-2055

[105] Setha S. Roles of abscisic acid in fruit ripening. Walailak Journal of Science and Technology. 2012;**9**:297-308

[106] Setha S, Kondo S, Hirai N, Ohigashi H. Quantification of ABA and its metabolites in sweet cherries using deuterium-labeled internal standards. Plant Growth Regulation. 2005;**45**:183-188. DOI: 10.1007/s10725-005-3088-7

[107] Rodrigo M-J, Alquezar B, Zacarías L. Cloning and characterization of two 9-cis-epoxycarotenoid dioxygenase genes, differentially regulated during fruit maturation and under stress conditions, from orange (Citrus sinensis L. Osbeck). Journal of Experimental Botany. 2006; **57**:633-643

[108] Rhodes MJC. The maturation and ripening of fruits. Senescence in Plants. 1980;**1980**:157-205

[109] Vendrell M, Palomer X. Hormonal control of fruit ripening in climacteric fruits. Acta Horticulturae. 1997;**463**:8

[110] Setha S, Kondo S, Hirai N, Ohigahi H. Xanthoxin, abscisic acid and its metabolite levels associated with apple fruit development. Plant Science. 2004; **166**:493-499

Chapter 5

# Role of Endophytes in Apple Replant Disease

*Ranjna Sharma, Joginder Pal, Deepika Sharma, Satish Kumar Sharma, Shalini Verma and Radhika Pathania*


## Abstract

Apple replant disease (ARD) is a major problem in all the apple-growing areas of the world. It is a complex problem. The exact cause of the problem is unknown, but soil biotic factors play a major role. The repeated cultivation of same crop on same land and exhaustion of nutrients of soil, persistence of soil-borne pathogens and changes in the pH of soil. Symptoms include stunting of tree growth with short internodes, small and light green rosette leaves, development of few lateral or feeder roots, underdeveloped root systems, decayed and discolored roots, poor establishment and severe disease results in the death of young trees and, sometimes, whole orchards. The endophytes provide direct benefits to host plants as they live in close proximity. Once they enter inside the host tissue they get easily established as they feel no competition with other microorganisms. Endophytes have the capacity to produce different secondary metabolites, which saves the host plants from biotic and abiotic stresses the host plants become resistant to both biotic and abiotic stresses. An interesting facet of the interaction between endophytes and their hosts is the capacity of many microorganisms to improve the plant's resistance by providing several bioactive metabolites. Therefore, the exploitation of soil microbial endophytes for the management of ARD is an important strategy.


**Keywords:** apple, replant disease, phytopathogens, endophytes, apple replant disease etiology

## 1. Introduction

The apple replant disease (ARD) is widespread throughout the world, and severity of this disease varies from region to region, fields, type of soil and presence of pathogenic microbial species in the rhizosphere. This problem may occur due to repeated cultivation of same crops on the same land and exhaustion of nutrients of soil, persistence of soil-borne pathogens and changes in the pH of soil. Replant disease is a debilitating soil problem affecting most orchards when they are replanted. This problem affects plants worldwide both in the nurseries, as well as in the orchards, in terms of plant growth, yield and quality of fruits [1, 2]. Different authors give different definitions of ARD. The main cause of apple replant disease is the imbalance in the soil microbial community [3], but the fungal pathogens vary from region to region. According to Utkhede [4], replant problem is caused by both biotic and abiotic factors, which suppress the growth of plants, whereas the replant disease is only caused by the biotic community present in the rhizospheric soil. Savory [5] defines ARD as soil sickness whose causes are unknown and uncertain. This problem is seen in other commercial crops, such as peach, pear, cherry, strawberry and rose.

Apple replant disease (ARD) affects the plant nurseries, as well as apple production worldwide by strongly reducing plant growth as well as fruit yield and quality [2, 3]. On ARD soils, over the lifetime of an apple orchard, a 50% reduced profitability has been estimated due to later and less fruit-bearing of the affected trees [6, 7]. ARD becomes problematic nowadays as the tree nurseries and fruit orchards are concentrated in certain regions of Pinneberg in Germany or Pistoia in Italy. Also, due to high-density plantations in apple crops, farmers are using dwarf rootstocks to achieve more yield in a short life span, which results in frequent replantation of apple orchards [8, 9]. Replant problem is very much severe and its control is very difficult. Crop rotation is the only method to reduce the effect of replantation, but it is also difficult due to alternative usage for industry, energy plants or other purposes. Different authors gave different definitions of the term 'replant disease' or related phrases, such as 'replant problem', 'soil sickness' or 'soil fatigue' [4, 10, 11]. These biotic and abiotic factors of replant problem suppress plant growth [4]. The unknown and uncertain cause of reduced growth is known as 'soil sickness' [5], which excludes nematode damage [12]. ARD is specifically related to species Malus domestica, and its persistence for decades is noticed [5]. The impact of ARD is reversible when plants are transplanted into virgin or healthy soil. This disease has been reported in other crops, such as rose, cherry, peach, strawberry and rowan, whereas roses are prone to it. Here we want to summarize the causes of ARD and their mitigation strategies. The apple trees may grow poorly when planted in non-sterilized soil. The problem occurs when same crop is planted in the same land year after a year, which disturbed the root microflora of the plants due to disturbed physiological and morphological reactions of apple plants to soil [13].

There are two forms of replant diseases: specific and non-specific. Specific replant disease only affects apple crops when the site is again replanted with the same crop, whereas non-specific replant disease occurs when stone fruits are planted on the site where previously planted with apple crops. The root cause of non-specific replant disease is nematode activity and fungal pathogens on the replanted sites. It was confirmed by using a nematicide pot test that 40% of Tasmanian apple orchards have both specific and non-specific replant disease (HAL project AP97005). It is a complex problem caused by number of pathogenic organisms, such as actinomycetes (filamentous bacteria), nematodes and bacteria, and fungi, such as *Rhizoctonia, Pythium* and *Phytophthora*. This has led to the conclusion that there is no specific treatment to control this disease with less impact on beneficial soil organisms. Soil sterilants, such as methyl bromide and chloropicrin, are used to kill the pathogenic microflora. Nowadays, there are eco-friendly approaches, such as organic matter, fertilizers, microorganisms, irrigation, cultivation and rootstocks, which have been used, but none of these are 100% effective to control the replant disease [14–16].

Apple replant disease is characterized by a severely reduced rate of root and shoot growth of second planting of same species or closely related species on same site, whereas non-specific causes of apple replant disease (ARD) mean incorrect use of fertilizers, poor soil structure, poor drainage, pH imbalances and presence of toxic compounds, such as herbicides, heavy metals and biological products. The main cause of ARD in New York apple replant orchards were nematodes (*Pratylenchus penetrans*), parasitic fungi, bacteria and other soil-borne microorganisms. The possible abiotic factors reported in the replant sites were poor nutrient availability, change in the pH of the soil, deteriorated soil structure and loss of organic matter, herbicide residues and other site-specific problems [17]. In Washington state soils, the *Pratylenchus penetrans* number was below the damage threshold level in eight of the nine orchards surveyed and bacteria were not identified in the disease. The main fungal pathogens associated with the disease were *Phytophthora cactorum, Pythium* spp., *Cylindrocarpan destructans* and *Rhizoctonia solani* [6].

Himachal Pradesh is known as 'apple state' of India as its cultivation has revolutionized the socio-economic status of the farmers and had played a pivotal role in the economy of growers. Apple orchards planted in late sixties have shown symptoms of declining productivity as these plants have completed their life span. Due to less land resources and choice of apple crop, old apple orchards are replanted with the same crop, which leads to drastic economic losses due to uprooting of old trees and poor establishment of new plantations on the replant site. The continuous cultivation of the same crop on the same field is the primary factor leading to replant problem. Due to this, a general decline in the growth and productivity of replanted apple orchards is commonly observed. The replant disease is one of the most important diseases in newly established orchards in old orchard sites of Himachal Pradesh [18]. This disease is very common in other stone fruit crops, such as cherry, strawberry, peach and pear, in Himachal Pradesh.

### 1.1 Symptoms

The disease symptoms of ARD are visible within 1 year of plantation. Disease symptoms affect the whole orchards, which include poor and uneven growth of trees and yield in future [10, 19]. It is very difficult to predict whether replant problem is present on a specific site or not. It is also difficult to assume that rootstock or nursery plants are the root cause for poor tree performance of apple trees. Replant disease affects most fruit crops including both pome and stone fruits.

Symptoms include stunting of tree growth with short internodes, small and light green rosette leaves, development of few lateral or feeder roots, underdeveloped root systems, decayed and discolored roots, poor establishment and severe disease results in the death of young trees and, sometimes, whole orchards. When a tree is uprooted, discolored roots, root tip necrosis and reduced root biomass can be seen. The cell wall degrading enzymes and cell-killing effector proteins, produced by pathogenic micro and macroflora, destroy the root tissues of host plants [20–24]. It impairs root function hampering plant responses to abiotic stresses, such as drought, flooding and nutrient deficiency.

### 1.2 Disease etiology

The apple replant disease is caused by number of pathogenic microflora and microfauna. The etiology of the disease varies from region to region and site to site. It depends on the dead tissues of the plant's remains in the soil, which insist the soil to attract the pathogenic microflora towards them. The ARD in Washington State is caused by different pathogenic fungi, such as *Rhizoctonia*, *Ilyonectria*; oomycetes: *Phytophthora*, *Pythium* and in some sites lesion nematode *Pratylenchus penetrans* [6]. The contaminated soil, irrigation water or planting stocks are different sources of inoculums to spread the ARD. The pathogenic fungi produce overwintering structures, such as sclerotia (*Rhizoctonia*), chlamydospores (*Ilyonectria*) and oospores (*Phytophthora* and *Pythium*), survive in dead or dormant roots. Lesion nematodes survive in the soil and plant residues as eggs and as multiple generations of juveniles and adults [25]. During the spring season, when new apple plants are planted on the same site, the active dormant roots secrete root exudates, which help the pathogen propagules to germinate and parasite nematode eggs to hatch. The propagules germinate to produce the fungal and oomycete mycelia and then infect the new fine roots. While fungi such as *Rhizoctonia* cause extensive root rot, *Pythium* spp. cause damping off in the root tissue resulting in diminished capacity of the plant to take up water and nutrients. The different causal agents of ARD, such as *Phytophthora cactorum*, Pythium ultimum, *Fusarium oxysporum and Dematophora necatrix,* were isolated from the temperate zones of Himachal Pradesh [18]. The primary cause of the disease is the over-exploitation of the old apple orchards specialized in fruit production. The causes of replant problem are not exactly clear but it is a complex problem and occurs mainly due to disturbances in the healthy microflora of soil, variation in soil pH, organic matter, root exudates of old tree roots, water retention in the rhizosphere area and plant genotypes [13].

Fungi of the genera *Cylindrocarpon, Rhizoctonia, Phytophthora* and *Pythium* are found frequently in ARD-affected soils and have proved to be crucial in the etiology of ARD [3, 18, 26]. Different *Fusarium* spp., such as *F. oxysporum, F. solani, F. equiseti* and *F. proliferatum,* were isolated from 10 different apple nurseries in Tunisia, which causes considerable losses to apple plants. Root endophytic *Cylindrocarpon*-like fungi *Thelonectria* sp. and *Ilyonectria* spp. were also associated with ARD [23], after *Pythium* spp., to be correlated to the growth reduction in the rootstock M9 growing in ARD-affected soil. Different species of Nectriaceae were also found in ARD-affected cortex cells applying laser microdissection [27]. Several fungal endophytes from ARD-affected apple roots were isolated and re-inoculated in a soil-free biotest [28].

Several *Streptomyces* amplicon sequence variants (ASVs) were detected in greenhouse ARD biotest, which were negatively correlated to shoot length and fresh mass, from both field sites. The *Streptomyces* ASVs in roots of apple plants in control soil increased their relative abundance over time. The 150 bacterial strains isolated by a culture-dependent approach revealed a high diversity of members of the genus *Pseudomonas*, confirming the data of the molecular barcoding approach. Therefore, it is important to combine these two approaches to better understand this complex disease and develop control measures. Finally, *Streptomyces* play a key role in the etiology of ARD in the study sites [29].

## 2. Mechanism of plant prophylaxis by endophytic microorganisms

The plant-associated microorganisms are important in reference to safety and quality of fruits. There are two types of soil microorganisms, rhizospheric and endophytic soil microorganisms. Both types of microorganisms have the same functions. The main difference between these two is, once the endophytic microorganisms enter the host tissue, they are protected from the changing environmental vagaries [30]. They provide direct benefits to host plants as they live in close proximity. Plant growth-promoting endophytic bacteria affect the plant growth through two mechanisms, such as direct and indirect ways [31]. The direct mechanisms include production of growth hormones (auxins, cytokinins and gibberellins), phosphate solubilization, siderophore production, competition and lytic enzymes secretions, whereas the indirect mechanisms of growth promotion include induction of plant resistance, predation and hyperparasite and production of antifungal metabolites, cell wall degrading enzymes, decreasing the amount of iron available to phytopathogens and synthesis of pathogen-inhibiting volatile compounds [32].

## 3. Exploitation of microbial endophytes as ARD management strategy

The word endophyte means 'in the plants'. These are isolated from internal tissues of plants by following the proper sterilization procedures. Microbial associations with host plants may be epiphytic, parasitic, mycorrhizal, endophytic, saprophytic etc. Only the epiphytic and endophytic microorganisms make their way to internal tissues of host plants. Endophytes include symbiotic associations of bacteria, fungi and yeast with the host plant [33]. Both partners are equally benefitted from each other. Many endophytes are members of common soil bacterial genera, such as *Bacillus, Pseudomonas, Burkholderia* and *Bacillus* [34]. Both the bacterial and fungal endophytes are present in the plant tissues without causing any ill effects. As the endophytes live inside the host tissue, they are beneficial to the host plants because they improve host plant tolerance to abiotic stresses, enhance growth, improve plant immune response and suppress pathogen colonization [35].

In comparison to rhizospheric microorganisms, endophytic microorganisms are more beneficial because they live in close proximity to host plants and exert direct benefit to host plants. Also, they live in non-competitive environments [36]. Endophytes improve plant growth by secreting phytohormones and consequently help in nutrition improvement using bidirectional nutrient transfer and enhancement of the health of plants by protecting them against phytopathogens [37, 38].

Endophytic bacteria are diverse in nature, most commonly represented by the genera *Pseudomonas, Bacillus, Burkholderia, Stenotrophomonas, Micrococcus, Pantoea* and *Microbacterium* [31]. The beneficial functions of well-known bacterial endophytic genera *Chitinophaga* and *Flavobacterium* in plant interactions in less known [39]. The most widely used genus, *Bacillus* has beneficial functions in its interactions with agriculturally-important plants. The endophytic bacteria indirectly promote plant growth by secreting antifungal metabolites or stimulating plant defense responses. The plant defense response, such as induced systemic responses (ISR), acts through specific plant response pathways, such as jasmonic acid (JA) pathway [40], but recently endophytic bacteria use the signalling pathways to stimulate these defense responses.

Endophytic fungi internally colonize plant tissues without causing any harm to the host plant [41]. The large group of endophytic fungi are commensals, some are mutualists [42, 43]. The most commonly exploited endophytic fungi belong to genera *Aspergillus*, *Bipolaris, Chaetomium, Cladosporium, Diaporthe, Fusarium, Alternaria, Mucor, Nigrospora, Paecilomyces, Penicillium, Piriformospora, Porostereum, Phoma, Trichoderma, Ulocladium and Yarrowia* [44, 45].

Therefore, the exploitation of these resident soil microbial communities for the management of ARD is a better strategy as plant-beneficial microflora helps in the establishment of new plants planted in old orchards sites and simultaneously increases the innate immunity of host plants by secreting bioactive metabolites. As we know that the bacteria on roots and in the rhizosphere is benefitted from plant root exudates. But it is not clear which population is more advantageous for the plants in terms of establishment of plants, growth, survival and yield. Endophytic populations like rhizospheric populations are conditioned by both the biotic and abiotic factors, but endophytes could be better protected from both the stress factors.

Exploitation of endophyte–plant associations result in the promotion of plant health and can play a significant role in low-input sustainable agriculture practices for all crops. By using molecular approaches, the whole genome sequences of key endophytic bacteria are available and the genes used in colonization and establishment of endophytic bacteria in plants can be identified.

*Trichoderma asperellum* strain 6S-2, an effective fungal endophyte isolated from roots of healthy apple trees growing in nine replanted orchards in Shandong Province, China, showed different lytic activities, such as protease, amylase, cellulase and laccase, which are important for the parasitic and antagonistic functions of pathogenic fungi. The inhibition rate of this fungal strain 6S-2 against phytopathogen *Fusarium proliferatum* f. sp. M. domestica MR5 was 52.41%. Strain 6S-2 also secreted different plant growth promoters, such as iron carriers, auxin and ammonia, and was able to solubilize phosphorus. The fermentation extract and volatile substances produced by endophyte *Trichoderma asperellum* inhibited the growth of *Fusarium proliferatum* f. sp. *M. domestica* MR5, by twisting, shrinking, swelling and rupturing the pathogen hyphae associated with apple replant disease [46].

Endophytic bacteria are able to lessen or prevent the deleterious effects of certain pathogenic organisms to host plants. The beneficial effects of bacterial endophytes on their host plant appear to occur through similar mechanisms as described for rhizosphere-associated bacteria. Bacterial endophytes may have an advantage over bacteria inhabiting the rhizosphere since living within a plant's tissues represents an opportunity to always be in contact with the plant's cells and therefore to more readily exert a direct beneficial effect. Of course, bacteria residing in the rhizosphere might also have the potential to enter and colonize the plant roots.

The maximum mycelial inhibition of 81.48% was obtained with the fungal endophyte *Aspergillus aculeatus* strain C2 under *in vitro* conditions. Microscopic studies on interaction between fungal endophytes with hyphal tips of apple root rot pathogen *Dematophora necatrix* revealed various morphological abnormalities in the hyphae like curling and bending of mycelium. Under glasshouse conditions, seed treatment pursued by soil application with fungal endophyte *Crinipellis tabtim* strain M8 isolate was highly effective and exhibited 93.55% disease control. Similarly, under field conditions, the overall maximum disease control was exhibited by *Crinipellis tabtim* strain M8 (84.95%). The most promising root endophytes were identified on morphological and ITS sequence analysis. Root colonization assay revealed maximum endosphere and rhizosphere colonization with *Crinipellis tabtim* strain M8. Also, confocal microscopic illustrations of transverse sections of root cells tenanted by fungal endophytes *Crinipellis tabtim* strain M8 as compared to untreated control suggested the persistence and establishment of endophytes in the endosphere of apple seedlings, shown in **Figure 1** [47].

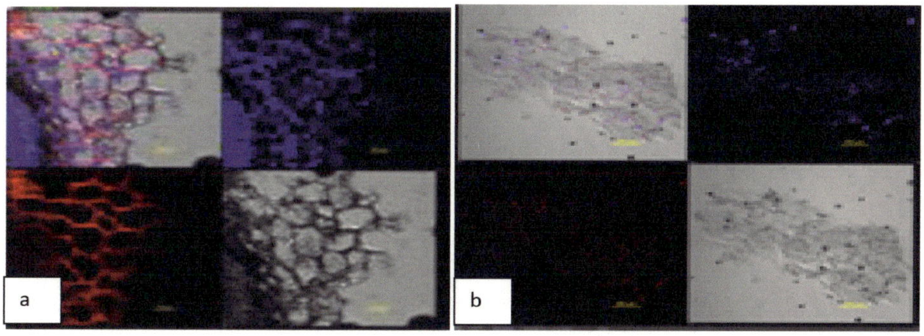

**Figure 1.**
*Microscopic elucidation of colonized fungal root endophytes. (a) Crinipellis tabtim strain M8. (b) Control by confocal laser scanning microscopy (CLSM by Pal et al. [47]).*

### 3.1 Helps in the establishment and proper growth of new apple plantation

Endophytes have the capacity to secrete different secondary metabolites, such as phenolic acids, alkaloids, quinones, steroids, saponins, tannins and terpenoids, which help the host plants to resist both biotic and abiotic stresses. The important secretions by these endophytes helped them in anticancer, antimalarial, antituberculosis, antiviral, antidiabetic, anti-inflammatory, anti-arthritis and immunosuppressive roles. The interaction between endophytes and their hosts is the capacity of many microorganisms to improve the plant's resistance by providing several bioactive metabolites [48]. The secretion of volatile compounds by plant-associated bacteria in association with host plants is a promising sustainable strategy to prevent the soil-borne and foliar fungal pathogens [49–51]. Plant-associated mutualistic microorganisms such as endophytic microorganisms, commonly known as endophytes, which colonize plants' internal tissues, frequently contribute to host metabolic function and protect plants against pests and diseases by producing biocontrol traits, such as bioactive secondary metabolites. The bacterial endophytes generally colonize the internal tissue of plants and are found nearly in every plant. The mechanism followed for plant growth promotion by some of the bacterial endophytes is same as that followed by rhizosphere bacteria. Inoculation of part of a plant with an endophyte may benefit plants via the production or suppression of phytohormones; for example, genes encoding proteins for biosynthesis of indole acetic acid (IAA), cytokinins (CKs) and gibberellins (GAs) are often present in the metagenome of plant endophytic bacterial communities. The application of spore suspension of *Trichoderma asperellum* strain 6S-2 to replanted apple orchard soils reduced plant oxidative damage and promoted plant growth in a pot experiment and the strain 6S-2 demonstrated plant growth-promoting activities such as protease, amylase, cellulase and laccase activities, which are important for the parasitic and antagonistic functions of pathogenic fungi. The inhibition rate of 6S-2 against *Fusarium proliferatum* f. sp. *M. domestica* MR5 was 52.41% and thus helps in the establishment of the young planatations to the replanted sites [13]. Compared to rhizosphere soil strains, endophytic microorganisms can effectively colonize host plants and better adapted to environmental changes, making them more effective in disease suppression and growth promotion [52]. The culture-independent and culture-dependent approach used in 3, 7 and 12 months planted apple plants in

ARD-affected and ARD un-affected soil at two sites reported that a high diversity of *Pseudomonas* in all soils and by using molecular bar-coding approaches an increase in relative abundance of *Actinobacteria* in plants grown in ARD and control plots [29].

### 3.2 Helps in disease suppression

Microbial endophytes live in association with the host plants without causing any side effects but their presence is beneficial to the host plant, as they protect the plants from biotic and abiotic stresses simultaneously enhance the growth and modulate plant immune response and suppress the pathogen colonization [35]. Endophytes reduce the negative impact of pathogens on their host by secreting siderophores, antibiotics, cell wall degrading enzymes, volatile organic compounds (VOCs), alkaloids, steroids, quinines, terpenoids, phenols and flavonoids, which are inhibitory to the phytopathogens [53, 54], or by interrupting the cross-talk signal of pathogens [55]. Secretion of lytic enzymes (β-1,3-glucanase, chitinase, cellulase and protease) by endophytes helps in degrading the cell wall of pathogenic fungi. Chitin, a major cell wall component of fungi, is degraded by enzyme chitinase. Chitinase produced by endophytic Streptomyces hygroscopicus, inhibit the growth of pathogenic fungi *Ralstonia solani*, *Fusarium oxysporum*, Alternaria alternata, *Aspergillus niger*, *Aspergillus flavus*, *Sclerotinia sclerotiorum*, *Hyaloperonospora parasitica* and *Botrytis cinerea* [56]. Endophytes help in plant growth either by producing secondary metabolite or nutrient assimilation or by preventing induction of plant disease symptoms by different pathogens. Endophytes produce a large number of bioactive compounds, which provide resistance to host plants and simultaneously protect the plants from biotic and abiotic stresses [57].

### 3.3 Increase the innate immunity of host plants

The bacterial endophytes enter the host system by following the same mechanisms as used by the bacterial pathogens, but the host plant allows the entry of specific genera to colonize the inner host tissues. Due to this, close interaction brings important changes in the plant physiology. The bacterial symbiont living in association with host exerts direct and indirect defenses to control plant biotic stresses. These defenses may be due to the secretion of volatiles and antibiotic compounds by endophytes, therefore, boosting the innate immunity of host plants against various diseases [58]. The plants allow the microbial colonization through phenotypic genes and producing metabolic signals. Host plants by evolving their genotypes provide sugar and lipids to the endophytes. Therefore, a plant's genotype can influence the microbiome composition and shape microbiome to enhance defense and mitigate the trade-off between growth and defense against pathogens by secreting chemoattractants [59].

## 4. Conclusions

Endophytes help in plant growth either by producing secondary metabolite or nutrient assimilation or by preventing induction of plant disease symptoms by different pathogens. The application of these endophytes may increase the availability of nutrients and control the replant disease organisms and considerably will regenerate, maintain and sustain the soil fertility and hence the establishment of apple rootstocks and yield in future. Therefore there is an urgent need for isolation, identification and characterization of indigenous plant growth promoting endophytes associated with apple plants as plant growth promoter in the fields as bioformulations. This will help to solve the replant problem of apple orchards and improve the economy of state.

## Acknowledgements

We are highly thankful to the Department of Science and Technology, GOI, New Delhi for providing financial assistance under the grant no. DST/WOS-B/2018/1833.

## Author details

Ranjna Sharma[1*], Joginder Pal[2], Deepika Sharma[3], Satish Kumar Sharma[3], Shalini Verma[3] and Radhika Pathania[4]

1 Department of Basic Sciences, Dr. YS Parmar University of Horticulture and Forestry, Solan, HP, India

2 Department of Plant Pathology, CSK, Palampur, HP, India

3 Department of Plant Pathology, Dr. YS Parmar University of Horticulture and Forestry, Solan, HP, India

4 Department of Environment Science, Dr. YS Parmar University of Horticulture and Forestry, Solan, HP, India

*Address all correspondence to: ranjnassharma@gmail.com

## References

[1] Manici LM, Kelderer M, Franke-Whittle IH, Rühmer T, Baab G, Nicoletti F, et al. Relationship between root-endophytic microbial communities and replant disease in specialized apple growing areas in Europe. Applied Soil Ecology. 2013;**72**:207-214. DOI: 10.1016/j. apsoil. 2013.07.011

[2] Mazzola M, Manici LM. Apple replant disease: Role of microbial ecology in cause and control. Annual Review of Phytopathology. 2012;**50**:45-65

[3] Manici LM, Kelderer M, Caputo F, Saccà ML, Nicoletti F, Topp AR, Mazzola M. Involvement of dactylonectria and Ilyonectria spp. in tree decline affecting multi-generation apple orchards. Plant and Soil. 2018;**425**:217-230. DOI: 10.1007/s11104-018-3571-3

[4] Utkhede RS. Soil sickness, replant problem or replant disease and its integrated control. Allelopathy Journal. 2006;**18**:23-38

[5] Savory BM. Specific replant diseases, causing root necrosis and growth depression in perennial fruit and plantation crops. Res Rev Commonw Bur Horticulture, East Malling. 1966;**1**:64

[6] Mazzola M. Elucidation of the microbial complex having a causal role in the development of apple replant disease in Washington. Phytopathology. 1998;**88**:930-938

[7] Van Schoor L, Sandra D, Cook NC. Characterizations of apple replant disease under South African conditions and potential biological management strategies. Scientia Horticulturae. 2009;**119**:153-162. DOI: 10.1016/j. scienta.2008.07.032

[8] St. Laurent A, Merwin IA, Fazio G, Thies JE, Brown MG. Rootstock genotype succession influences apple replant disease and root-zone microbial community composition in an orchard soil. Plant and Soil. 2010;**337**:259-72. DOI: 10.1007/s11104-010-0522

[9] Volk GM, Chao CT, Norelli J, Brown SK, Fazio G, Peace C, et al. The vulnerability of US apple (Malus) genetic resources. Genetic Resources and Crop Evolution. 2015;**62**:765-794. DOI: 10.1007/s10722-014-0194-2

[10] Hoestra H. Replant Diseases of Apple in the Netherlands. Dissertation Landbouwhogesch: Wageningen; 1968. Available from: https://edepot.wur.nl/285366

[11] Klaus H. Das Problem der BodenmüdigkeitunterBerücksichtigung des Obstbaus. Landw Jahrb. 1939;**89**: 413-459

[12] Spethmann W, Otto G. Replant problems and soil sickness. In: Roberts AV, Debener T, Gudin S, editors. Encyclopedia of Rose Science. Oxford, UK: Elsevier LTD; 2003. pp. 169-180

[13] Winkelmann T, Smalla K, Amelung W, Baab G, Grunewaldt-Stöcker G, Kanfra X, et al. Apple replant disease: Causes and mitigation strategies. Current Issues in Molecular Biology. 2019;**30**:89-106

[14] Mazzola M. Identification and pathogenicity of Rhizoctonia spp. isolated from apple roots and orchard soils. Phytopathology. 1997;**87**:582-587. DOI: 10.1094/PHYTO.1997.87.6.582

[15] Wilson S, Andrews P, Nair TS. Non-fumigant management of apple

replant disease. Scientia Horticulturae. 2004;**102**:221-231

[16] Rumberger A, Shengrui Y, Merwin IA, Nelson EB, Thies JE. Rootstock genotype and orchard replant position rather than soil fumigation or compost amendment determine tree growth and rhizosphere bacterial community composition in an apple replant soil. Plant and Soil. 2004;**264**:247-60

[17] Mai WF, Merwin IA, Abawi GS. Diagnosis, etiology and management of replant disorders in New York cherry and apple orchards. Acta Horticulturae. 1994;**363**:33-42. DOI: 10.17660/ActaHortic.1994.363.5

[18] Sharma R, Pal J, Rana S, Kaur M. Suppression of soil borne fungal pathogens associated with apple replant disease by cyclic application of native strains of *Pseudomonas aeruginosa*. Journal of Applied and Natural Science. 2017;**9**:2105-2109

[19] Szczygiel A, Zepp AL. An occurrence and importance of apple replant problem in Polish orchards. ActaHorticulturae. 1998;**77**:99-102

[20] Grabke A, Hu MJ, Luo CX, Bryson PK, Schnabel G. First report of brown rot of apple caused by *Moniliniafructicola*, in Germany. Plant Disease. 2011;**95**:772. DOI: 10.1094/PDIS-02-11-0113

[21] Isman MB. Plant essential oils for pest and disease management. Crop Protection. 2000;**19**:603-608. DOI: 10.1016/S0261-2194(00)00079-X

[22] Lobet G, Pages L, Draye X. A novel image-analysis toolbox enabling quantitative analysis of root system architecture. Plant Physiology. 2011;**157**:29-39. DOI: 10.1104/pp.111.179895

[23] Mazzola M, Brown J, Zhao X, Izzo AD, Fazio G. Interaction of brassicaceous seed meal and apple rootstock on recovery of *Pythium*spp and *Pratylenchuspenetrans* from roots grown in replant soils. Plant Disease. 2008;**93**:51-57. DOI: 10.1094/PDIS-93-1-0051

[24] Mazzola M, Gu YH. Impact of wheat cultivation on microbial communities from replant soils and apple growth in greenhouse trials. Phytopathology. 2000;**90**:114-119. DOI: 10.1094/PHYTO.2000.90.2.114

[25] Agrios G. Plant Pathology. Burlington, MA: Elsevier Academic Press; 2005. p. 948

[26] Kelderer M, Manici LM, Caputo F, Thalheimer M. Planting in the 'inter-row' to overcome replant disease in apple orchards: A study on the effectiveness of the practice based on microbial indicators. Plant and Soil. 2012;**357**:381-393

[27] Popp C, Wamhoff D, Winkelmann T, Maiss E, Grunewaldt SG. Molecular identification of Nectriaceae in infections of apple replant disease affected roots collected by Harris Uni-Core punching or laser microdissection. Journal of Plant Diseases. 2020;**127**:571-582. DOI: 10.1007/s41348-020-00333-x

[28] Popp C, Grunewaldt-Stocker G, Maiss E. A soil free method for assessing pathogenicity of fungal isolates from apple roots. Journal of Plant Diseases and Protection. 2019;**126**:329-341. DOI: 10.1007/s41348-019-00236-6

[29] Mahnkopp-Dirks F, Radl V, Kublik S, Gschwendtner S, Schloter M,

Winkelmann T. Dynamics of bacterial root endophytes of *Malusdomestica* plants grown in field soils affected by apple replant disease. Frontiers in Microbiology. 2022;**13**:841558. DOI: 10.3389/fmicb.2022.841558

[30] Glick BR. Plant growth-promoting bacteria: Mechanisms and applications. Scientifica. 2012;**2012**:963401

[31] Santoyo G, Moreno-Hagelsieb G, Carmen Orozco-Mosqueda M, Glick BR. Plant growth-promoting bacterial endophytes. Microbiological Research. 2016;**183**:92-99. DOI: 10.1016/j.micres.2015.11.008

[32] Glick BR. Beneficial Plant-Bacterial Interactions. Springer, Heidelberg; 2015. pp. 65-96

[33] Brader G, Compant S, Mitter B, Trognitz F, Sessitsch A. Metabolic potential of endophytic bacteria. Current Opinion in Biotechnology. 2014;**27**:30-37

[34] Lodewyckx C, Mergeay M, Vangronsveld J, Clijsters H. Van der Lelie, D. Isolation, characterization and identification of bacteria associated to the zinc hyperaccumulator *Thlaspicaerulescens* subsp. *calaminaria*. International Journal of Phytoremediation. 2002;**4**:101-115

[35] Dini-Andreote F. Endophytes: The second layer of plant defense. Trends in Plant Science. 2020;**25**:319-322. DOI: 10.1016/j.tplants.2020.01.007

[36] Compant S, Clement C, Sessitsch A. Plant growth–promoting bacteria in the rhizo- and endosphere' of plants: Their role, colonization, mechanisms involved and prospects for utilization. Soil Biology and Biochemistry. 2010;**42**:669-678

[37] Andreozzi A, Prieto P, Mercado-Blanco J, Monaco S, Zampieri E, Romano S. Efficient colonization of the endophytes *Herbaspirillumhuttiense* RCA24 and *Enterobacter cloacae* RCA25 influences the physiological parameters of *Oryza sativa* L. cv. *Baldo rice*. Environmental Microbiology. 2019;**21**:3489-3504. DOI: 10.1111/1462-2920.14688

[38] Shen FT, Yen JH, Liao CS, Chen WC, Chao YT. Screening of rice endophyticbiofertilizers with fungicide tolerance and plant growth-promoting characteristics. Sustainability. 2019;**11**:1133. DOI: 10.3390/su11041133

[39] Carrion VJ, Perez-Jaramillo J, Cordovez V, Tracanna V, MDE H, Ruiz-Buck D. Pathogen-induced activation of disease-suppressive functions in the endophytic root microbiome. Science. 2019;**366**:606-612. DOI: 10.1126/science.aaw9285

[40] Montejano-Ramírez V, García-Pineda E, Valencia-Cantero E. Bacterial compound N, N-dimethylhexadecylamine modulates expression of Iron deficiency and defense response genes in Medicagotruncatula independently of the jasmonic acid pathway. Plants. 2020;**9**:624. DOI: 10.3390/plants9050624

[41] Hallmann J, Quadt-Hallmann A, Mahaffee WF, Kloepper JW. Bacterialendophytes in agricultural crops. Canadian Journal of Microbiology. 1997;**43**:895-914. DOI: 10.1139/m97-131

[42] Afzal I, Shinwari ZK, Sikandar S, Shahzad S. Plant beneficial endophytic bacteria: Mechanisms, diversity, host range and genetic determinants. Microbiological Research. 2019;**221**:36-49. DOI: 10.1016/j.micres.2019.02.001

[43] Bacon CW, White J, editors. Microbial Endophytes. CRC Press; 2000

[44] Omomowo OI, Babalola OO. Bacterial and fungal endophytes: Tiny giants withimmense beneficial potential for plant growth and sustainable agricultural productivity. Microorganisms. 2019;7:481. DOI: 10.3390/7110481

[45] Vu T, Sikora R, Hauschild R. *Fusarium oxysporum* endophytes induced systemicresistance against *Radopholussimilis* on banana. Nematology. 2006;**8**:847-852. DOI: 10.1163/156854106779799259

[46] Wang H, Zhang R, Duan Y, Jiang W, Chen X, Shen X, et al. The endophytic strain *Trichodermaasperellum* 6S-2: An efficient bio-control agent against apple replant disease in China and a potential plant-growth-promoting fungus. Journal of Fungi. 2021;7:1050. DOI: 10.3390/jof7121050

[47] Pal J, Sharma SK, Devi S, Sharma R, Raj H, Karn M, et al. Screening, identification, and colonization of fungal root endophytes against *Dematophora necatrix*: A ubiquitous pathogen of fruit trees. Egyptian Journal of Biological Pest Control. 2020;**30**:112. DOI: 10.1186/s41938-020-00312-2

[48] Gunatilaka AAL. Natural products from plant-associated microorganisms: Distribution, structural diversity, bioactivity and implications of their occurrence. Journal of Natural Products. 2006;**69**:509-526

[49] Bailly A, Weisskopf L. The modulating effect of bacterial volatiles on plant growth: Current knowledge and future challenges. Plant Signaling & Behavior. 2012;7:79-85. DOI: 10.4161/psb.7.1.18418

[50] Garbeva P, Weisskopf L. Airborne medicine: Bacterial volatiles and their influence on plant health. The New Phytologist. 2020;**226**:32-43. DOI: 10.1111/nph.16282

[51] Koberl M, Ramadan EM, Adam M, Cardinale M, Hallmann J, Heuer H. Bacillus and Streptomyces were selected as broad-spectrum antagonists against soil-borne pathogens from arid areas in Egypt. FEMS Microbiology Letters. 2013;**342**:168-178. DOI: 10.111/1574-6968.12089

[52] Bogner CW, Kariuki GM, Elashry A, Sichtermann G, Buch AK, Mishra B, et al. Fungal root endophytes of tomato from Kenya and their nematode biocontrol potential. Mycological Progress. 2016;**15**:30

[53] Chung JH, Song GC, Ryu CM. Sweet scents from good bacteria: Case studies on bacterial volatile compounds for plant growth and immunity. Plant Molecular Biology. 2016;**90**:677-687. DOI: 10.1007/s11103-015-0344-8

[54] Ek-Ramos MJ, Gomez-Flores R, Orozco-Flores AA, Rodríguez-Padilla C, González-Ochoa G, Tamez-Guerra P. Bioactive products from plant-endophytic gram-positive bacteria. Frontiers in Microbiology. 2019;**10**:463. DOI: 10.3389/fmicb.2019.00463

[55] Miller MB, Bassler BL. Quorum sensing in bacteria. Annual Review of Microbiology. 2001;**55**:165-199. DOI: 10.1146/annurev.micro.55.1.165

[56] Haggag WM, Abdallh EG. Purification and characterization of chitinase produced by endophytic *Streptomyces hygroscopicus* against some phytopathogens. Microbiological Research. 2012;**2**:145-151. DOI: 10.5923/j.microbiology.20120205.06

[57] Fadiji AE, Babalola OO. Elucidating mechanisms of endophytes used in plant protection and other bioactivities

with multifunctional prospects. Frontiers in Bioengineering and Biotechnology. 2020;**8**:467. DOI: 10.3389/fbioe.2020.00467

[58] Oukala N, Aissat K, Pastor V. Bacterial endophytes: The hidden actor in plant immune responses against biotic stress. Plants(Basel). 2021;**10**:1012. DOI: 10.3390/plants10051012

[59] Lu H, Wei T, Lou H, Shu X, Chen Q. A critical review on communication mechanism within plant-endophytic fungi interactions to cope with biotic and abiotic stresses. Journal of Fungi (Basel). 2021;**7**:719. DOI: 10.3390/jof7090719

Chapter 6

# The Role of Major Phenolics in Apple to Total Antioxidant Capacity

*Sweta Kumari, Swati Manohar, Priyanjali Kumari, Veda Krishnan, Chirag Maheshwari, Sneh Narwal, Om Prakash Gupta, Vinutha T. Gowda, Navita Bansal and Anil Dahuja*

## Abstract

The naturally occurring phenolic compounds have received major attention in recent years as huge amounts of phenolic compounds can be extracted from fruits, vegetables and beverages that have substantial health benefits. From a physiological and metabolic aspect, phenolic compounds are vital in defence responses, such as anti-ageing, anti-inflammatory, anti-oxidant and anti-proliferative, anti-bacterial, anti-hyperlipidemic, anti-cancer, anti-diabetic, neuroprotective, cardioprotective activities. Among the fruits having a higher content of phenolic compounds, the apple (*Malus Domestica*) is the most widely consumed fruit in the world. Apples have a high nutritional value as it is a rich source of ascorbic acid, polyphenols and pectin. Apple peel forms a small percentage (6–8%) of the total fruit weight and contains the highest content of phenolic compounds, particularly chlorogenic acid. There are five major groups of polyphenolic compounds found in apples namely flavanols (Catechin, Epicatechin and Pyrocyanidins), phenolic compounds, phenolic acids (mainly Chlorogenic acids), dihydrochalcones (Phloretin glycosides), flavonols (Quercetin glycosides) and anthocyanins (Cyanidin). This chapter reviews the chemical properties, mode of action, types, extraction of phenolics in apples and the contribution and role of major phenolics in apples to the total antioxidant capacity.

**Keywords:** phenolic compound, flavanols, flavonols, dihydrochalcones, antioxidant capacity, anti-hyperlipidemic, contribution

## 1. Introduction

Metabolic functions are one of the important characteristics of living organisms. From a simple prokaryotic cell to a highly specialised eukaryotic organism, proper functioning of metabolism is essential for the survival of the organism. These metabolic processes are important for the survival, growth and dispersal of plants as well. In simple terms, metabolism refers to all the chemical processes in the body that convert food into energy. This energy is made available to all other cellular processes essential to sustaining life. Metabolism is further of two types-primary metabolism and secondary metabolism. Primary metabolism is responsible for the proper functioning of all major physiological processes of plants and animals while secondary metabolism includes all other metabolic pathways that are not essential for survival but plays various other important roles. Both types of metabolic processes result in the formation of certain chemical compounds known as metabolites. Primary metabolites are formed as a result of primary metabolic processes while secondary metabolism results in the formation of secondary metabolites.

*Secondary plant metabolites* are specialised chemical compounds produced by plants and do not possess any role in primary metabolic processes but are essential for the survival of plants in their environment. Earlier it was believed that secondary metabolites are formed as a by-product of primary metabolic reactions that are of no significance to plant growth. However, researchers have revealed various important functions of secondary plant metabolites. Though they are not directly involved in essential physiological processes, affect and regulate these processes by acting as intermediates in these processes. Secondary metabolic compounds in plants can be broadly grouped into 4 groups (**Figure 1**). Terpenes, Phenolics, nitrogen-containing compounds and sulphur-containing compounds. Each secondary metabolite has a specific role to perform (**Figure 2**). They act as structural components of plants in the form of lignins. They also act as a defence against pathogens and herbivores. Secondary metabolites provide better tolerance to stress conditions and protect against UV rays. Certain secondary metabolite act as an attracting agents for pollination and seed dispersal. They also inhibit the growth of microbes and check the competition by inhibiting the growth of nearby plants. Various research has established the fact that secondary metabolites act as bioactive compounds in fruits and vegetables.

*Phenolics* are one of the most significant phytochemicals which are naturally produced as a secondary metabolite in plants. Plant phenolics are bioactive compounds that have been reported to exhibit various biochemical properties such as antioxidant, structural polymer (lignin), attractants (flavonoids and carotenoids), UV

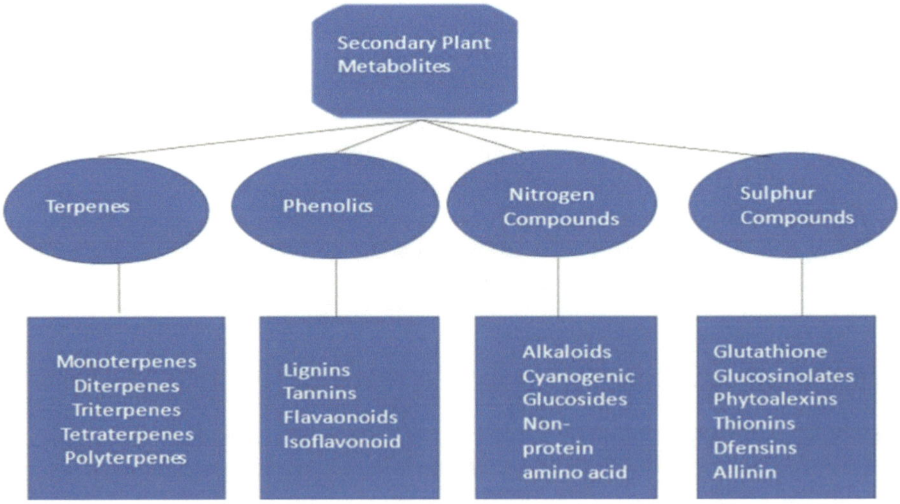


**Figure 1.**
*Types of metabolites.*

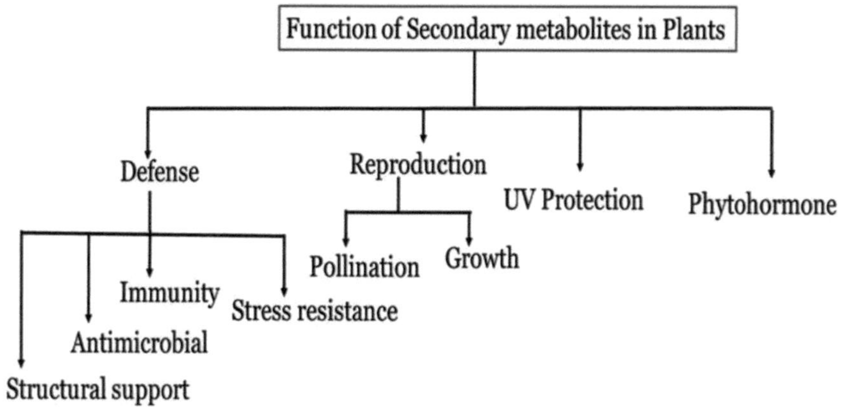


**Figure 2.**
*Functions of secondary metabolites.*

screens(flavonoids), signal compounds (salicylic acid and flavonoids) and defence response chemicals (tannins and phytoalexins) [1].

Chemically, phenolics are compounds that have an aromatic ring or a benzene ring with one or more hydroxide groups. They may be in the form of simple phenols or polyphenols based on the number of phenol or hydroxide groups present on them and also possess variation in the number of attached carbon atoms (**Table 1**). They may also possess functional derivatives such as esters, methyl esters, glycosides *etc.* attached to them.

In plants, they are found in conjugated form and may be joined to carboxylic acids, organic acids, amines, lipids and other phenolic compounds. Phenolics play important roles in plants as they protect plants from harmful ultraviolet radiation, pathogen and parasite infection as well as from predators. A phenolic compound such as Anthocyanin provides red, orange, blue or purple colour to fruits.

### 1.1 Mechanism of action of phenolic compounds

Phenolic compounds are bioactive in nature and act as anti-oxidant by reacting with a variety of free radicals. The mechanism of anti-oxidant actions involves either hydrogen atom transfer or transfer of a single electron or sequential proton loss electron transfer or by chelation of transition metals [2].

| No. of C atoms | Basic structure | Class |
|---|---|---|
| 6 | $C_6$ | Simple Phenols; Benzoquinones |
| 7 | $C_6$-$C_1$ | Phenolic acids |
| 8 | $C_6$-$C_2$ | Acetophenone; Phenylacetic acid |
| 9 | $C_6$-$C_3$ | Hdroxycinnamic acid; Coumarin; Isocoumarin |
| 10 | $C_6$-$C_4$ | Napthoquinone |
| 15 | $C_6$-$C_3$-$C_6$ | Flavonoids; Isoflavonoids |
| 30 | $(C_6\text{-}C_3\text{-}C_6)_2$ | Biflavonoids |
| n | $(C_6\text{-}C_3)_n$ $(C_6)_n$ $(C_6\text{-}C_3\text{-}C_6)_n$ | Lignins; Catecholmelamine; |

**Table 1.**
*Types of phenolics based on number of carbon atoms.*

### 1.2 Antioxidant

One of the most important properties of phenolics is their antioxidant property which has become a matter of interest for the scientific community in recent years. *Antioxidants* are substances that inhibit the process of oxidation and thus prevent or delay the damage to the cells caused due to free radicals [3] or unstable molecules that are produced in the body as a response to biological, metabolic, environmental or other factors. Free radicals are waste substances produced by cells as the body processes food and reacts to the environment. Antioxidants are supposed to be" free radical scavengers" as they help in neutralising free radicals in our bodies and in this way benefit our health. Several factors influence the activity or effectiveness of antioxidants. These factors include the chemical structure of the antioxidant, its concentration, temperature and its location in the system.

Fruits and vegetables are rich sources of antioxidants that enhance the nutritional quality of fruits and vegetables. The antioxidant property of phenols present in fruits and vegetables is beneficial to human health as researchers have suggested that antioxidants can lower the risk of various chronic diseases such as cancer, heart stroke and age-related macular degeneration [4].

## 2. Apple

Apple (*Malus Domestica*) is the fruit of a domesticated tree belonging to the genus Malus and the family Rosaceae. Apples are one of the most widely cultivated tree fruits. Apples originated in central Asia and have been grown for thousands of years in Asia and Europe and were brought to North America by Europeans. From sweet red varieties like Red Delicious, Fuji or Gala to tangy green ones like Granny Smith, over 7500 different cultivars of apples are available worldwide that makes apples also the most widely consumed fruit globally [5].

According to data from Food and Agriculture Organisation Corporate Statistical Database, total apple production in 2017 was 83,139,326 metric tonnes and in 2020, it was 87,236,221 metric tonnes. China tops the list of the highest apple-producing countries in the world followed by the United States and Turkey [6]. Based on a comparison of 161 countries in 2019, Hungary ranked first in apple consumption per capita with 33.3 kg followed by the Netherlands and Albania [7].

Apart from higher production, several other factors make apples the most widely consumed fruit. These factors include easy market availability, cost affordability, long shelf life, variety of processed apple products such as Jams, pies, canned apples, apple juice, smoothies etc.

There is nothing new to talk about the health benefits of eating apples as the whole world is familiar with a very popular saying "An apple a day keeps the doctor away." Apples are incredibly nutritious fruits [8] (**Table 2**). Apples are low in sodium, fat and cholesterol. They are a very good source of Vitamin C, fibre and antioxidants.

### 2.1 Antioxidant capacity of apples

The antioxidant activity of apples is mainly attributed to the phenolic compounds present in apples. There is a correlation between the phenolic content of apples and antioxidant activity. The apple varieties with higher phenolics tend to have higher antioxidant activity. Various phenolic compounds are present in different parts of

| **Amount per 100 grams** | |
|---|---|
| Calories | 52 cal. |
| Total Carbohydrate | 14 g |
| Total Fat | 0.2 g |
| Protein | 0.3 g |
| Cholesterol | 0 mg |
| Calcium | 0% |
| Magnesium | 1% |
| Sodium | 1 mg |
| Potassium | 107 mg |
| Dietary Fibre | 2.4 g |
| Sugar | 10 g |
| Vitamin C | 7% |
| Vitamin B6 | 0% |
| Vitamin D | 0% |

*Sources: United States Department of Agriculture (per cent daily values are based on a 2000 calorie diet) [6].*

**Table 2.**
*Nutrition value of apple.*

apples such as peels, pulp, core, and seeds in different concentrations (**Figure 3**). Generally, apple peel is rich in the concentration of phenolic compounds as compared to flesh or pulp part. Apples had the highest portion of free phenolics when compared to other fruits which makes them easily absorbable into the bloodstream.

In the United States, 22 per cent of the phenolics consumed from fruits are from apples making them the largest source of phenolics [9]. When compared to many fruits in the United States, apples had the second highest level of antioxidant capacity [10].

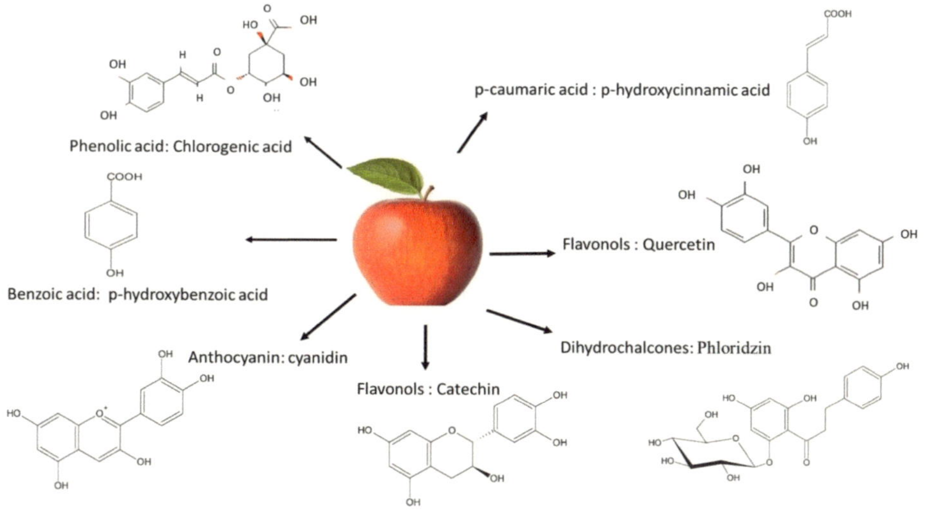


**Figure 3.**
*Enriched phenolic compounds in apples.*

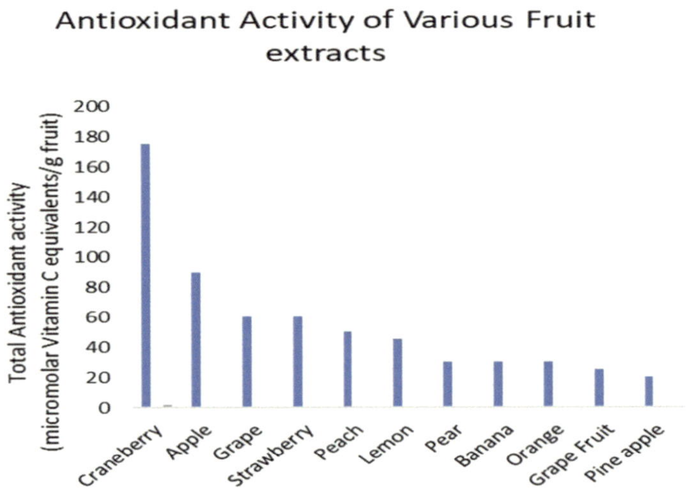


**Figure 4.**
*Chart depicting the antioxidant activity of different fruits.*

According to available literature and research, antioxidant activities and total phenolic content of 62 fruits (**Figure 4**) were evaluated using ferric reducing anti-oxidant power (FRAP) and Trolox equivalent antioxidant capacity (TEAC) assay and the Folin-Ciocalteu methods. The experiment revealed statistical data related to different varieties of apples which showed that apples belonging to the cultivars of Green-delicious, Red-delicious and Rose-red had intermediate values of 68.29, 73.96 and 70.57 mg GAE/100 g of wet weight respectively [11].

### 2.2 Flesh vs. peel

Certain individual phenolic compounds may also be found in higher concentrations in flesh as compared to peel. "Chlorogenic acid" is a type of phenolic compound found in apples that were found in peel in higher concentration as compared to flesh in apple varieties of Elstar, Fuji, McIntosh Pinova, Red Delicious, but Idared, Golden Delicious, Granny Smith, Reineta possessed higher concentration of chlorogenic acid in flesh than peel [12]. Another type of phenolic compound known as Anthocyanin which is responsible for the bright red skin colouration of apples is highly concentrated in the peel. The peel of Granny Smith, Golden Delicious, and McIntosh is richer in p-coumaroylquinic acid content while in Gala, the flesh has a higher content of the same phenolic compound [12].

### 2.3 Phenolic content in apple seeds

Apple seeds are also a good source of polyphenols particularly phloridzin. Apart from phloridzin, other phenolic compounds are also found in apple seeds such as dihydrochalcones, hydroxycinnamic acid, hyperin, chlorogenic acid, quercetin, caffeic acid, protocatechuic acid, and flavan-3-ols which are found in monomeric and oligomeric forms as well as polymeric forms. Monomeric forms include Catechin and Epicatechin while oligomeric forms such as proanthocyanin are present. Flavonols are also found in apple seeds.

HPLC analysis has proved phloridzin as the chief phenolic compound found in seeds as phloridzin content was found to be 240.45–864.42 mg/100gDW which was measured by the Folin-Ciocalieu assay of apple seeds of seven cultivars. The seven apple cultivars included varieties of Gale Gala, Starking, Honeycrisp, Fuji, Qinguan, Golden Delicious, and Qinyang. Among the seven cultivars, Honeycrisp showed the highest phenolic content while Qinyang showed the lowest value of phenolic compounds [13].

## 3. Factors affecting phenolic content of apples

The difference in the content of individual phenolic compounds in apples of different varieties can be attributed to different biosynthesis pathways of phenolic compounds. Other factors affecting the phenolic concentration in apples include cultivar, maturity of the fruit, conditions of cultivation, growing season, harvest, storage, and environmental factors. Comparative study of antioxidant activity and phenolic content of apples is difficult to evaluate due to different extraction methods, different cultivars of apples, geographical region, soil type, and light hours. It is also supposed that organically grown apples have higher phenolic content as compared to the apples grown by integrated farming. The genetic constitution of the different cultivars of apples majorly affects the phenolic content in apples. One important factor that affects the phenolic content of apples is the ripening or maturation condition of apples. Researchers have found that the amount of dihydrochalcones in very young apples is very high but after 14 weeks the number of dihydrochalcones, flavonols, and chlorogenic acid decreased in both the peel as well as the flesh. Various studies regarding the apple varieties and factors affecting the phenolic content have revealed a general trend in the phenolic content and its concentration in different parts of apples. The antioxidant properties of different parts of apples follow the order: peel>core>flesh. This has also been revealed that the apple peel contains two to six times more phenolic compounds as compared to the flesh of the apple [12].

## 4. Biosynthesis pathway of phenolics in apples

Biosynthesis of phenolic compounds may occur via various pathways (**Figure 5.**) i.e., the Shikimate pathway, Phenylpropanoid pathway, and Flavonoid pathways. The

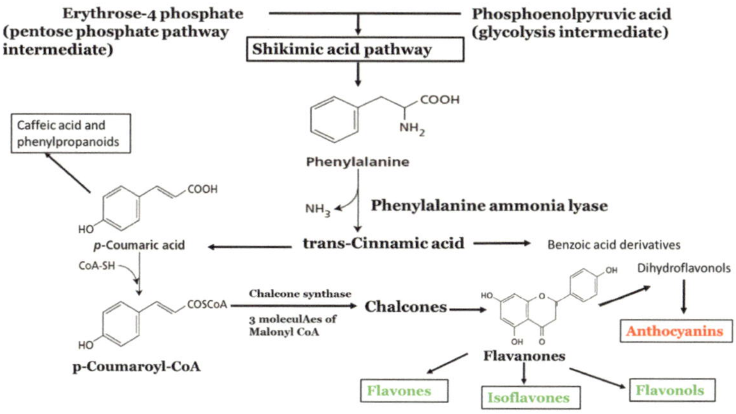


**Figure 5.**
*Biosynthesis pathway of phenolics in apple.*

first step in the synthesis of phenolic compounds is the involvement of glucose in the Pentose Phosphate Pathway (PPP) and the transformation of glucose-6-phosphate into ribulose-5-phosphate. The first step in conversion to ribulose-5-phosphate is catalysed by the enzyme glucose-6-dehydrogenase. The conversion to ribulose-5-phosphate produces an equivalent amount of nicotinamide adenine dinucleotide phosphate (NADPH) which acts as a reducing agent for cellular anabolic reactions.

PPP also produces erythrose-4-phosphate along with phosphoenolpyruvate from glycolysis. This phosphoenolpyruvate is used in the Phenylpropanoid pathway to produce phenolic compounds after going through the Shikimic pathway in which phenylalanine is produced.

## 5. Extraction of phenolic compounds from apples

The extraction of phenolic compounds from apples is a very crucial step in the study and characterisation of chemical and nutraceutical properties of phenolic compounds as only after proper extraction the phenolic compound could be studied and used as a dietary supplement, cosmetic product or pharmaceutical. Earlier Conventional methods such as maceration, decoction, and soxhlet were used for the extraction of phenolic compounds from apples. These days, non-conventional methods have overtaken conventional methods and are being extensively used for the extraction of phenolics. The non-conventional methods include UAE or ultrasound-assisted extraction, MAE or microwave-assisted extraction, SFE or supercritical fluid extraction, PLE or pressurised liquid extraction, and ASE or accelerated solvent extraction. The non-conventional techniques are better than conventional methods concerning yield, time, cost, solvent saving and other factors. Generally, before extraction, the samples have to go through various procedures such as milling, grinding, homogenisation etc. other procedures include air drying or freeze drying. Freeze drying is prefered over air drying because it retains the phenolic content of the sample to the maximum extent. The most commonly used method is extracting phenolic compounds using a suitable solvent. The generally used solvents for extraction procedures are methanol, ethanol, acetone, and ethyl acetate. The yield of the phenolic compound using the chemical method varies with the type of solvent used. Usually, phenolic compounds having low molecular weight give better yield in methanol while the polyphenols having higher molecular weight such as flavanols give better yield with acetone.

### 5.1 Introduction to different non-conventional methods of extraction

*Ultrasound-assisted extraction or UAE* is a non-conventional extraction technique where the extraction procedure is assisted by ultrasound waves. The efficiency of this technique greatly depends on cell disruption and effective mass transfer.

In *Microwave assisted extraction or MAE*, solvents in contact with the samples are heated using microwave energy. This feature of fastly heating the sample solvent is the main advantage of using this technique. MAE technique takes about 15 to 30 minutes but uses a very small amount of sample volume as compared to conventional methods of techniques.

In the *Supercritical fluid extraction or SFE* technique, the extractant that is the phenolic compound is separated from the matrix using supercritical fluids as extracting solvents. The main advantages of using this technique are its selectivity and high speed as the extraction process is based on the diffusion of the solvent into the matrix. The matrix used for extraction is generally solid but also may be in liquid form. One disadvantage of the SFE technique is the requirement of high pressure for extraction which makes this technique costlier as compared to other conventional methods of extraction.

*Accelerated Solvent extraction and Pressurised liquid extraction or PLE* techniques are quite similar to the supercritical fluid extraction technique in the respect that it also uses a solid or semi-solid matrix for extraction and also require a high degree of temperature and pressure.

### 5.2 Antioxidant activity measurement assay

Nowadays sophisticated methods are available for the antioxidant activity measurement assay. Chemical methods with sensitive and automated detection features are being used for the assay of antioxidant activity. The experiment for assay of the antioxidant activity was performed on two cultivars of apple namely the Idared and the Sampion. The antioxidant activity was estimated using ABTS and DPPH assays. The polyphenol profile was determined by the HPLC method. Seeds of the sample apple cultivars showed higher antioxidant capacity and also a higher content of phenolic compounds found as compared to their peel and flesh. The two phenolic compounds found in abundance in seed and peel respectively were Phloridzin and Quercetin glycoside. The result of the assay may be seen in **Table 3** [14].

### 5.3 Antioxidant activity assay techniques

*Radical scavenging method (ROS)* requires no lipids and uses a chemical system containing an oxidant, an oxidizable probe and antioxidants under assay. The mechanism followed may be either hydrogen atom transfer or electron transfer.

*Oxygen radical absorbance capacity (ORAC) assay* works by monitoring inhibition of peroxyl radical-induced oxidation.ORAC can detect both lipophilic as well as hydrophilic antioxidants.

*Chemiluminiescence* method of antioxidant assay requires a chemiluminescence species, an oxidant generally used is hydrogen peroxide and the extract to be detected. It is a very sensitive and low-cost tool used for the antioxidant assay.

*DPPH radical scavenging* is a commonly used antioxidant assay tool frequently used for the antioxidant assay of apples. Chemically, DPPH is 2,2-Diphenyl-1-picrylhydrazyl and a stable chromogen and deep purple in colour. DPPH method of assay is based on an electron transfer mechanism also hydrogen atom transfer being an intermediate reaction pathway.

*Trolox equivalent antioxidant capacity (TEAC) assay* measures the total antioxidant capacity by the ability of antioxidants to scavenge ABTS which is a stable radical. Chemically ABTS is (2,2′-azinobis(3-ethylbenzothiazoline-6-sulphonic acid)). TEAC

| Cultivar | Part | Phenolic compound | Percentage of compound |
|---|---|---|---|
| Idared | Seed | Phloridzin | 84 |
| Sampion | Seed | Phloridzin | 72 |
| Idared | Peel | Quercetin glycoside | 54 |
| Sampion | Peel | Quercetin glycoside | 38 |

**Table 3.**
*Phenolic content: part vs. variety.*

also employs electron transfer and hydrogen atom transfer mechanisms. When combined with HPLC, it provides an efficient rapid and accurate method for the detection of individual compounds.

*Ferric reducing antioxidant power (FRAP)* is an electron transfer-based method which uses ferric trivalent cation as an oxidant. It is a very simple and low-cost method of antioxidant assay.

*Total phenolic content (TPC)* is an important tool for measuring total antioxidant activity, particularly in the case of apples. The Folin-Ciocalteu assay is generally used for the estimation of Total phenolic content. The assay uses the Folin-Ciocalteu reagent which is reduced by phenolic compounds under alkaline conditions. It is a very simple and reliable method for the assay of total antioxidant capacity but has a few drawbacks also. Folin-Ciocalteu assay is sensitive to ph, and temperature changes, therefore proper care must be taken for the accurate assay of antioxidant capacity while using this technique.

### 5.4 Estimation protocol of phenolic compounds

The commonly used methods for the estimation of different phenolic compounds are HPLC combined with Folin-Ciocalteu assay and chromatography analysis. Samples extracted from different varieties of fruits are collected with a mixture of acetone/water to achieve a good yield of polyphenol content. The polyphenols are analysed by normal phase HPLC and LC-MS. The total phenolic content of the sample is measured using the Folin-Ciocalteu assay. The different phenolic compounds present in the sample are detected by chromatographic analysis.

## 6. Phytochemical profile of apples and their bioavailability

The apple is a very nutritious food and has a very rich phytochemical profile, particularly that of phenolic compounds. Various types of phenolic compounds are found in all parts of the apple including the seeds and core. The phytochemical profile of different varieties of apples differs in terms of the content of phenolic compound and its amount also varies in different parts of apples. Phenolics are the most significant phytochemical found in apples. The phenolic content of apples includes many flavonoid compounds, phenolic acids, majorly chlorogenic acid and also hydroxycinnamic acid in small quantities, quercetin glycosides, phloretin glycosides and anthocyanins. The phenolic content of apples may vary from 0.1 g/Kg fresh weight to 10 g/Kg in certain Cider apples [15]. The apple cultivar Honeycrisp shows the highest phenolic content while the Qinyang variety showed the least content of phenolic compounds. The apple cultivar Granny Smith showed higher content of p-coumaroylquinic acid in the peel while the same phenolic content was higher in the flesh of the Gala cultivar.

Any chemical compound that is a part of our diet and beneficial to our body benefits only when it is absorbed to the highest extent in our systemic circulation. The term "*Bioavailability*" refers to the rate and extent of absorption of a particular chemical compound by the body of the organism consuming that compound. The higher the bioavailability of a compound, the greater its effectiveness. Apples have a very rich content of beneficial chemical compounds and possess high bioavailability of various phenolic compounds.

### 6.1 Factors affecting the bioavailability of phenolic compounds in apples

The bioavailability of different phenolic compounds found in apples is not the same. Many factors affect the bioavailability of different phenolics in apples. These factors may be Environmental factors, Food processing techniques, Chemical composition of food and Type of phenolic compound.

Environmental factors include sunlight duration, ripening stage of apple, rainfall, fruit yield etc. food processing techniques may affect the bioavailability of apple phenolics by increasing or decreasing the phenolic content of food. The different food processing techniques include Thermal treatment, Homogenisation, cooking method and method of storage. Raw apples have a higher content of phenolic compounds while cooked apples show lower phenolic content because excessive heating of apples during cooking reduces phenolic content. Homogenisation may cause alteration of the apple matrix thereby increasing the bioavailability of its phenolic content. Food processing techniques are also an important factor affecting apples' bioavailability of phenolic content. It is generally found that apple puree and apple juice have higher bioavailability as compared to unprocessed apples. That is because the phenolic content is easily and quickly absorbed from the processed apple. The method and duration of storage of apples and their products also affect the bioavailability of apple phenolics. The presence and absence of positive or negative effects in our diet may also affect the bioavailability of phenolic content from apples. The storage of apple juices for 11 months resulted in a decrease in phenolic content in apple juice [16].

### 6.2 Major phenolics found in apples

The two broad groups of phenolic compounds found in apples are the Flavonoids and Phenolic acids with each group consisting of a large number of compounds (**Table 4**) with varying structures.

Flavonoids are a class of polyphenolic secondary metabolites found in plants and commonly consumed in the diet of humans. Chemically, flavonoids have a general structure of a 15-carbon skeleton which consists of two phenyl rings and a heterocyclic ring. The carbon structure can be abbreviated as C6-C3-C6.

Flavonoids may be further grouped into classes such as flavanols (Catechin, Epicatechin, Pyrocyanidins), dihydrochalcones (Phloridizin, Phloretin glycosides), flavonols (Quercetin glycosides), anthocyanins.

Flavanols are derivatives of flavans that possess a 2-phenyl-3,4-dihydro-2H-chromen-3-of skeleton. Flavan-3-ol has 2 chiral carbons. Catechin and Epicatechin are epicatechins, with (−) epicatechin and (+) catechin being the most optical isomers found in nature.

Dihydrochalcones or 1,3-Diphenylpropan-1-one are organic compounds with the formula C6H5C(CH2)2C6H5 and a molar mass of 210.27 g/mol. It is a white solid and soluble in many organic solvents. Phenolic compound Phloretin belongs to this group which occurs as glycosides in apples.

Anthocyanins are glycosides of anthocyanidins, that are water-soluble vacuolar pigments. In acidic conditions, anthocyanins appear as red pigments while in alkaline conditions, they appear as blue pigments. They are abundantly present in apple peels of red apples.

The term Phenolic acids refer to the phenolic compounds having one carboxylic acid group. They are generally found in bound form as amides, esters, and glycosides and rarely occur in free form. Phenolic acids are mainly divided into two subgroups

| Phenolics types | Phytochemicals |
|---|---|
| Phenol | Chlorogenic acid<br>Hydroxy benzoic acid<br>Hydroxycinnamic acid |
| Flavanoids | |
| Anthocyanidins | Cyanidin 3-O-arabinose<br>Cyanidin 3-O-galactoside<br>Cyanidin 3-O-xyloside<br>Cyanidin 3-O-xylgalactoside |
| Flavonols | Quercetin 3-arabinopyranoside<br>Quercetin 3-arabinofuranoside<br>Quercetin 3-galactoside<br>Quercetin 3-glucoside<br>Quercetin 3-rhamnoside<br>Quercetin 3-rutinoside<br>Quercetin 3-xyloside |
| Dihydrochalcones | Phloretin<br>Phloretin-20-O-xyloglucoside<br>Phloridzin<br>3-Hydroxyphloridzin |
| Flavan-3-ols | Monomeric; (+) – Catechin, Epicatechin<br>Oligomeric (Procyanidins)<br>Procyanidin B1<br>Procyanidin B2<br>Procyanidin B5<br>Procyanidin B7<br>Procyanidin C1 |

**Table 4.**
*Potential phenolics in apple.*

hydroxybenzoic and hydroxycinnamic acid. The most abundant soluble bound hydroxycinnamic acid present in apples is Chlorogenic acid.

Chlorogenic acid is the ester of caffeic acid and quinic acid functioning as an intermediate in lignin biosynthesis. The acid has the formula $C_{16}H_{18}O_9$ and a molar mass of 354.31 g/mol.

### 6.3 Role of phenolics

Phenolics play an important role in plant development as phenolic compounds are significant molecules for the biosynthesis of lignin and other plant pigments. In plants, phenolic compounds have anti-microbial activity and also resist pathogens.

## 7. Apple: eating an apple a day keeps the harmful radicals away!

The tremendous contribution of phenolics and flavonoids to antioxidant capacity provides a significant role as a scavenger and alleviates the harmful effect of reactive oxygen species in the cell. Research has suggested that the various polyphenol compounds of apples are effective in preventing several chronic diseases such as heart disease, cancer etc.

Polyphenols also exhibit anti-obesity effects. Flavanoids such as epicatechin may prevent heart disease by lowering blood pressure, reducing LDL cholesterol and reducing atherosclerosis [17].

Quercetin and phloridzin reduce type 2 diabetes risk. Quercetin's anti-inflammatory effects reduce insulin resistance while phloridzin cuts sugar uptake by the intestines [18].

Quercetin helps in regulating the immune system by reducing inflammation. It is also effective against bronchial asthma and sinusitis [19].

Proanthocyanidins have been reported to prevent and reduce allergic asthma airway inflammation [20].

Various researchers have also suggested that phenolic compounds have anti-cancer properties and are effective in preventing certain cancers, i.e., lung breast and digestive tract cancer [21].

Chlorogenic acid is supposed to inhibit weight gain, prevent liver steatosis and block insulin resistance [22].

Various researchers have reported that apple extracts also show antimicrobial properties. According to literature data, extracts of the peel of the Royal Gala variety and the Granny Smith variety showed inhibition of the growth of *Escherichia coli*, *Pseudomonas aeruginosa* and staphylococcus aureus strains.

## 8. The role of phenolics in apples to the total antioxidant capacity

Research that aimed to examine the average concentrations of the major phenolic compounds in six cultivars of apples. The average phenolic concentration detected among six cultivars is summarised in the table below and a pie chart (**Figure 6**) (derived from **Table 5**) showing percentages of different phenolics [23].

The results derived from high-performance liquid chromatography with diode array detection [24] performed on different varieties of apples are summarised in the chart (**Table 6**).

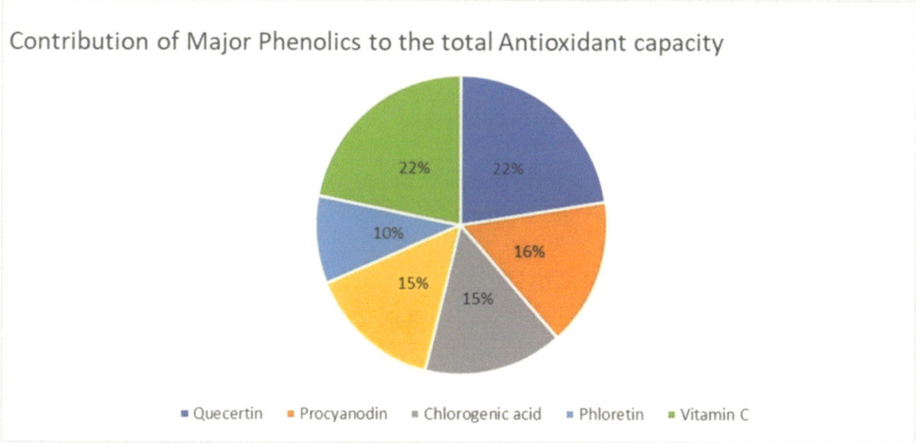


**Figure 6.**
*Pie chart depicting the contribution of different phenolics to the total antioxidant capacity.*

| Compound detected | Amount present (mg/100 g) |
|---|---|
| Quercetin glycosides | 13.2 |
| Procyanidin B | 9.35 |
| Chlorogenic acid | 9.02 |
| Epicatechin | 8.65 |
| Phloretin glycosides | 5.59 |
| Vitamin C | 12.8 |

**Table 5.**
*Amount of different phenolic compounds in apples.*

| Apple variety | Part of apple | Content of major phenolic compound | Antioxidant contribution | Benefits | Reference |
|---|---|---|---|---|---|
| Golden delicious | Peel | Catechin | 66–164 | Prevent cell damage | [24] |
| Red delicious | Peel | Phloridzin | 104–159 | Anti-diabetic | [24] |
| Reineta | Peel | Epicatechin | 283–439 | Anti-hypertensive | [24] |
| Red delicious | peel | Chlorogenic acid | 113–157 | Anti-obesity | [24] |

**Table 6.**
*The major content of phenolic compounds with their antioxidant contribution (mg/kg fresh weight).*

### 8.1 Phenolic compounds to the total antioxidant capacity of apples

According to available literature and experiments conducted by various researchers, we may conclude that flavanol content contributes most to the antioxidant capacity of apples. The table and pie chart also supports that the antioxidant activity of apples is majorly due to phenolic compounds which are flavanoids and phenolic acids and Vitamin C has an insignificant contribution to the antioxidant capacity of apple as compared to phenolic compounds.

## 9. Phenolics: emerging nutraceutical

Phenolic compounds obtained naturally from plants have lately become significant molecules. Scientists hold high aspirations as the research suggests phenolics to be a compound of high nutraceutical value which could be used in the diet, drugs cosmetics etc. many researchers have indicated the promising result of the use of phenolic compounds in the industrial sector. A very well-researched property of Phenolic compounds is their antioxidant property which has shown significant results in various health ailments. Phenolic compounds can be used in food as additives and preservatives; in cosmetics as UV protection and anti-ageing agents. Research has given hopes of using phenolic compounds as potential pharmaceuticals which may in future provide cures to many incurable and chronic diseases. However, there is still a lack of knowledge and awareness about the bioavailability and nutritional value of phenolic compounds and further research is being done. The potential use of phenolic compounds in different spheres of life may become a substitute for many artificially synthesised compounds and would prove to be a boon for the human race.

## 10. Conclusion

Apples have been praised for their health benefits for ages and now it is evident as per the research and experiments conducted on the quantitative and qualitative characterisation of apple phytochemicals. The different parts of apples such as peel, flesh and even seeds are enriched with several types of phenolic compounds. Comprehensive research is being done by the scientific community for the advancement of techniques used for the extraction and processing of phenolic compounds from apples so that the bioavailability of the phenolic compounds could be enhanced. The higher concentration of phenolic compounds in apples has drawn the attention of the health sector due to its potential to be used as a dietary supplement as well as an essential component of many industrial products such as drugs and cosmetics. Several phenolic compounds have been detected in the different parts and different cultivars of apples and each with specific health benefits. The amount of phenolic compounds present in apples is higher than the other phytochemicals in apples. The antioxidant activity of apples due to vitamin C and other phytochemicals is very low as compared to the antioxidant activity due to the concentration of the phenolic compound in apples. According to this review based on the available literature data, it is concluded that the antioxidant capacity of apples is significantly due to the phenolic compounds present in them. The apple is a very good source of antioxidants and should be consumed with peel as the peel shows a higher level of antioxidant activity.

## Conflict of interest

The authors declare no conflict of interest.

## Author details

Sweta Kumari[1*], Swati Manohar[2*], Priyanjali Kumari[2], Veda Krishnan[1], Chirag Maheshwari[1], Sneh Narwal[1], Om Prakash Gupta[3], Vinutha T. Gowda[1], Navita Bansal[1] and Anil Dahuja[1*]

1 Division of Biochemistry, ICAR-Indian Agricultural Research Institute, New Delhi, India

2 Department of Biotechnology, School of Life Sciences, Mahatma Gandhi Central University, Motihari, India

3 Quality and Basic Science, Indian Institute of Wheat and Barley Research, Karnal, Haryana, India

*Address all correspondence to: sweta381@gmail.com; swatiee.25@gmail.com and ad_bio@yahoo.com

## References

[1] Lin D, Xiao M, Zhao J, Li Z, Xing B, Li X, et al. An overview of plant phenolic compounds and their importance in human nutrition and management of type 2 diabetes. Molecules. 2016;**21**(10):1374. DOI: 10.3390/molecules21101374

[2] Zeb A. Concept, mechanism, and applications of phenolic antioxidants in foods. Journal of Food Biochemistry. 2020;**44**(9):1-22. DOI: 10.1111/jfbc.13394

[3] Pham-Huy LA, He H, Pham-Huy C. Free radicals, antioxidants in diseases and health. International Journal of Biomedical Sciences. 2008;**4**(2):89-96

[4] Pawlowska E, Szczepanska J, Blasiak J. Dietary polyphenols in age-related macular degeneration: Protection against oxidative stress and beyond. Oxidative Medicine and Cellular Longevity. 2019;**2019**:1-13. DOI: 10.1155/2019/9682318

[5] Elzebroek ATG, Wind K. Guide to Cultivated Plants. Wallington: CABI; 2008

[6] Report: Fresh Deciduous Fruits Annual. 2016. United States Department of Agriculture. Available from: www.ers.usda.gov

[7] Food and Agriculture Organisation Corporate Statistics Database. 2022. Available from: www.hegilibrary.com

[8] United States Department of Agriculture. 2019. Available from: www.fdc.nal.usda.gov

[9] Vinson JA, Xuehui S, Zubik L, Bose P. Antioxidant quantity and quality in foods: Fruits. Journal of Agricultural Food Chemistry. 2001;**49**:5319. DOI: 10.1021/jf0009293

[10] Boyer J, Liu RH. Apple phytochemicals and their health benefits. Nutrition Journal. 2004;**3**(5):1-15. DOI: 10.1186/1475-2891-3-5

[11] Fu L, Xu BT, Xu XR, Gan RY, Zhang Y, Xia EQ, et al. Antioxidant capacities and total phenolic contents of 62 fruits. Food Chemistry. 2011;**129**:345-350. DOI: 10.1016/j.foodchem.2011.04.079

[12] Kalinowska M, Bielawska A, Siwkiewicz HL, Priebe W, Lewandowski W. Apples: The content of phenolic compounds vs. variety, part of apple and cultivation model, extraction of phenolic compounds, biological properties. Plant Physiology and Biochemistry. 2014;**84**:169-188. DOI: 10.1016/j.plantphy.2014.09.006

[13] Ying X, Fan M, Ran J, Zhang T, Sun H, Dong M, et al. Variation in phenolic compounds and antioxidant activity in apple seeds of seven cultivars. Saudi Journal of Biological Science. 2016;**23**:379-388. DOI: 10.1016/j.sjbs.2015.04.002

[14] Duda-Chodak A, Tarko T, Tuszynski T. Antioxidant activity of apples—An impact of maturity stage and fruit part. Acta Scientiarum Polonorum, Technologia Ailmentaria. 2011;**10**(4):443-454

[15] Sanoner P, Guyot S, Marnet N, Molle D, Drilleau JP. Polyphenol profiles of french cider apple varieties (Malus domestica sp.). Journal of Agricultural and Food Chemistry. 1919;**47**(12):4847-4853. DOI: 10.1021/jf990563y

[16] Gliszczynska-Swiglo A, Tyrakowska B. Quality of commercial apple juices evaluated on the basis of the polyphenol content and TEAC. Journal of Food Science. 2003;**68**(5):1844-1849. DOI: 10.1111/j.1365-2621. 2003.tb12340

[17] Parmenter BH, Croft KD, Hodgson JM, Dalgaard F, Bondonno CP, Lewis JR, et al. An overview and update on the epidemiology of flavonoid intake and cardiovascular disease risk. Food Function. 2020;**11**:6777-6806. DOI: 10.1039/DOFO01118E

[18] Niederberger KE, Tennant DR, Bellion P. Dietary intake of phloridzin from a natural occurrence. British Journal of Nutrition. 2020;**123**:942-950. DOI: 10.1017/s0007114520000033

[19] Mlcek J, Jurikova T, Skrovankova S, Sochor J. Quercetin and its anti-allergic immune response. Molecules. 2016;**21**(5): 623. DOI: 10.3390/molecules 21050623

[20] Coleman SL, Shaw OM. Progress in the understanding of the pathology of allergic asthma and the potential of fruit proanthocyanidins as modulators of airway inflammation. Food Function. 2017;**8**(12):4315-4324. DOI: 10.1039/c7fo00789b

[21] Fabiani R, Minelli L, Rosignoli P. Apple intake and cancer risk: A systematic review and meta-analysis of observational studies. Public Health Nutrition. 2015;**19**(14):2603-2617. DOI: 10.1017/s136898001600032x

[22] Santana-GálvezJ, Cisneros-ZevallosL, Jacobo-Velázquez DA. Chlorogenic acid: Recent advances on its dual role as a food additive and a nutraceutical against metabolic syndrome. Molecules: A Journal of Synthetic Chemistry and Natural Product Chemistry. 2017;**22**(3):358. DOI: 10.3390/molecules 22030358

[23] Lee KW, Kim J, Kim D-O, Lee HJ, Lee CY. Major phenolics in apple and their contribution to total antioxidant capacity. A Journal of Agricultural Food Chemistry. 2003;**51**(22):6516-6520. DOI: 10.1021/jf034475w

[24] Escarpa A, Gonzalez MC. High-performance liquid chromatography with diode array detection for the determination of phenolic compounds in peel and pulp from different apple varieties. Journal of Chromatography A. 1998;**823**:331-337. DOI: 10.1016/s0021-9673(98)00294-5

Section 3

# Cultivars, Rootstocks, Nursery and Production Techniques of Malus Species

Chapter 7

# Apple Production under Protective Netting Systems

*Richard M. Bastías and Alexandra Boini*

## Abstract

Apple crop is more and more cultivated under protective netting systems. Depending on the location and sunlight intensity, apple orchards can benefit from these installations, as they will be protected against extreme weather events. Depending on the technical features of the thread, the nets will be hail-proof, wind-proof, or rain-proof, while having different shading percentages. Modern fruit production faces high pressure also related to biotic stressors; thus, modern protective nets are designed to aid pest management. These protective systems become interesting, as they will induce changes in the orchards' microenvironment, with consequences on crop physiology. Netting mainly reduces incoming solar radiation and wind speed, altering the heat balance. Leaf gas exchanges and water relations can be positively influenced by netting in apple cultivation areas with extreme solar radiation, high temperatures, and low water availability. These considerations are important, especially if the final yield and quality are not compromised by shading. These protective systems can allow higher sustainability of apple production, lowering resource use, along with crop protection.

**Keywords:** *Malus domestica* borkh, protected crop, nets, sunlight, orchard management, physiology

## 1. Introduction

Orchard netting is a technique that has become widespread in apple production to prevent damage from adverse climatic events, such as sunburn in countries such as Chile, South Africa, the United States, and Australia [1–4], hail in countries such as Germany, Italy, and Spain [5–7], and insect attacks such as codling moth in countries such as Italy, France, and Canada [6, 8, 9]. This wide range of uses of netting in apple trees implies the adoption of different installation systems and net designs, differentially impacting microclimatic conditions of light, temperature, relative humidity, and wind speed [9–11], and with a consequent effect on plant physiological responses, such as leaf gas exchange, water relations, tree growth, floral development, and fruit quality traits [12–16]. Greater environmental and biological stress pressure due to climate change is forcing the use of netting to be essential by apple growers, but at the same time imposes the challenge of adjusting orchard management practices to the particular microclimate and physiological plant conditions that are generated under netting, such as irrigation, pruning, crop load regulation, and pest control [7, 17], as well as the need for innovation in new netting systems that are compatible with sustainable fruit production [18]. This chapter provides an overview of netting in apple orchards, including installation systems, net designs, their effects on microclimate and plant physiology, and the innovative development of photo-selective nets and sustainable netting systems.

## 2. Netting systems

### 2.1 Netting structure

The most appropriate netting structure for apple orchard covering depends on the commercial purpose, installation cost, climatic conditions, and benefits pursued [19]. The most common structure system is the anti-hail roof type (**Figure 1** left-above), which has become widespread among apple growers in countries such as Italy, Germany, and Spain [20, 21], and some South American countries such as Chile and Argentina [17]. This system allows good ventilation of the orchard and, thanks to the slope of the roof, favors the discharge of hail that slides and accumulates toward the net junction between the rows, thus avoiding damage to the installation system due to overweight from an extreme hailstorm [19–21]. A second system is the shading roof system (**Figure 1** right-above), which is widely used in apple production areas with a high incidence of solar radiation and extreme temperatures, and to protect fruits against sun damage [17]. It has the advantage of greater access to machinery between rows and lower investment costs in netting materials and structure while being

**Figure 1.**
*Apple orchards under netting structures of anti-hail roof system (left-above), shading roof system (right-above), plot exclusion system (left-below), and single row exclusion system (right-below).*

also effective in mitigating the effects of hail and wind. However, the disadvantage of this system is the low stability against extreme hailstorms, given the absence of slopes, thus not allowing hail to slide and discharge, increasing the structure tension by the hail weight [22]. Other systems are exclusion nets that can totally plot the orchard, or just single rows (**Figure 1** left-right-below). These systems offer total protection against hail, sunburn, and the impact of wind [22]. Additionally, they allow protection from the attack of pests, limiting chemical pesticide use. It has been demonstrated that the single-row exclusion system in apple orchards allows a reduction of fruit injury by codling moths near to 99%, without any application of specific insecticide, since this netting excludes the male moths flying over the tree canopy [8]. However, this type of system requires ventilation management, especially in areas with hot summers, since they can induce problems of increased temperature and relative humidity. Furthermore, these systems are more expensive, since more netting materials are needed to cover the total plots or single rows of apple orchards [21, 22]. Besides, it has been observed that these systems tend to accumulate dust on the topmost part, year after year, along with moss formation.

### 2.2 Net designs

Different types of nets are used in apple orchards, which differ in their weaving and colors. The first is the "Raschel" type net, which is commonly used for shading, with flat threads joined by chains of transverse threads, called the warp, that tie the transverse flat threads, called weft (**Figure 2**), preventing them from falling apart due to the action of the wind or hail fall [20]. The second group of nets is the "Leno" type net (**Figure 1**), which is made of monofilament yarns, woven orthogonally in both weft and warp directions. The warp corresponds to a double fiber of threads that transversally encloses each thread of the weft that is positioned longitudinally (**Figure 2**). This type of net is more rigid than the previous one and is more suitable against strong hailstorms [20]. From the mechanical point of view, the "Raschel" net exhibits a very fragile and linear elastic tension in the weft direction, and higher resistance and nonlinear tension in the warp direction, while the "Leno" anti-hail net shows fragile and lineal tension in the warp direction, and higher strength nonlinear

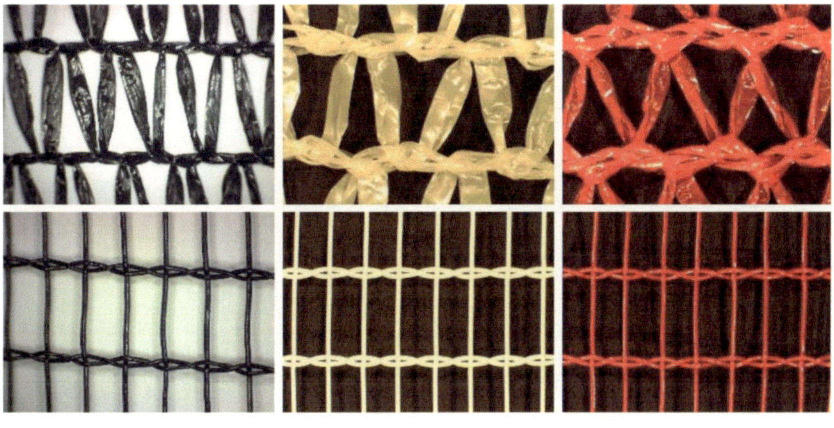

**Figure 2.**
*Details of threads design of black, white, and red shade nets (above), and black, white, and red hail net (below) commonly used to protect apple orchards.*

tension in the weft direction [23]. The color of the net is obtained by adding pigments to the HDPE polymer during the manufacture of the yarns. The most common colors of nets used in apple orchards are black, white, and red (**Figure 2**), although blue, green, yellow, and gray nets also have been used by apple growers. The color of the net exerts a selective effect on sunlight transmission, altering the light spectrum and the ratio of direct vs. diffuse light. This will ensure the effects on the plant's physiological and productive responses as has been widely evaluated in apple trees and will be analyzed in more detail during the development of this chapter [10–15].

## 3. Microclimate conditions

### 3.1 Light conditions

The main microclimate impact of netting is in the reduction of incoming photosynthetic active radiation (PAR), which could be affecting the most important physiological process determining the fruit yield and quality in apples, such as photosynthesis and carbon allocation [24]. The influence on nets in the PAR reduction is widely affected by the net design; thus, translucid-type nets reduce up to 7% of PAR light availability, while PAR under black anti-hail is reduced by 18%. In addition, the combination of thread colors also influences the quantity of PAR light reduction under netting. White-green net reduces up to 13% of PAR transmission, while under red-black nets the reduction of PAR availability is near 16% [25]. These differences in PAR availability among nets are due to the fact that the color of the threads alters the diffuse light proportion concerning total light transmitted under netting. In this sense, it has been shown that apple orchards covered with a pearl net provide 4% more PAR light available than those covered with a red net, due to the increase in diffuse light [26]. In this case, the pearl net increases the proportion of diffuse light by up to 15% compared to the red net (**Figure 3**).

On the other hand, the color of nets has a direct influence on the spectra transmission of sunlight [11, 27]. White, gray, or black color nets do not alter the spectral transmission of light at any of the wavelengths (**Figure 4**), while red and blue netting differentially alter light transmission in the specific wavelengths. The red net reduces

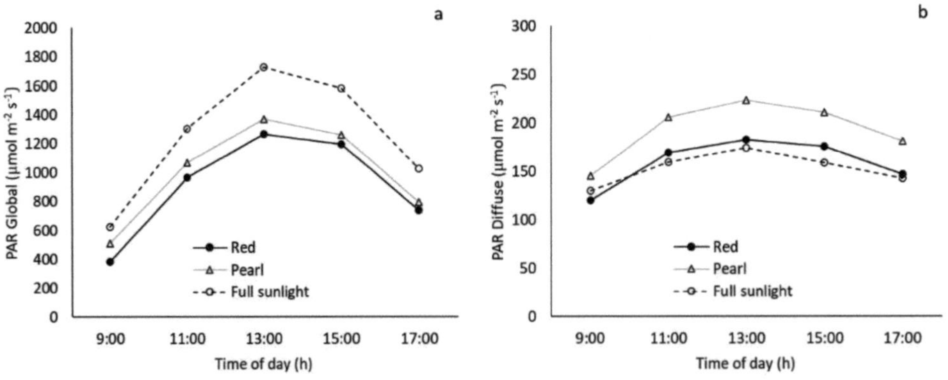


**Figure 3.**
*Total (a) and diffuse (b) photosynthetic light incident in "Gala" apple orchard under pearl and red netting (adapted from Umanzor et al., 2017).*

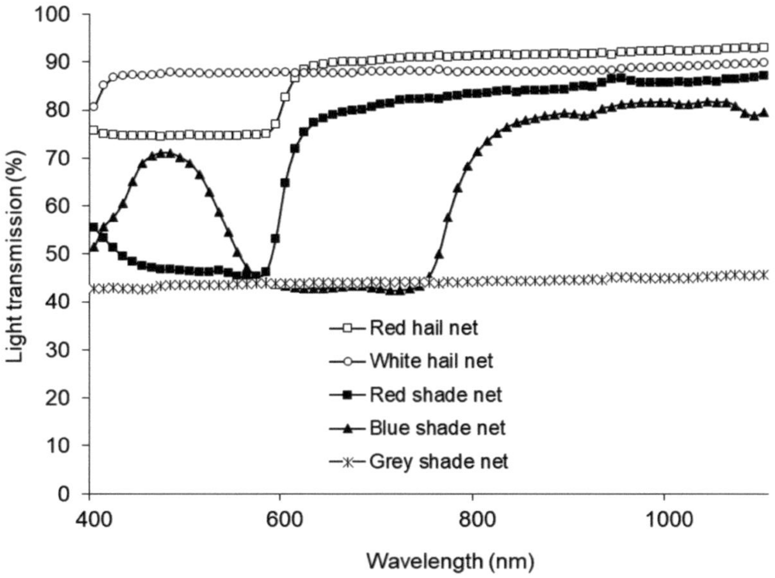


**Figure 4.**
*Spectra light transmission of colored hail and shade nets commonly used to protect apple orchards (adapted from Bastías et al., 2012; 2021).*

transmission in the blue light spectrum (400–500 nm) and increases it in the red (600–700 nm) and far-red (700–800 nm) light spectrum, whereas the blue net alters light transmission in the opposite way (**Figure 4**). It has been shown that these changes in the wavelength of light transmission modify different vegetative and reproductive responses in apple orchards grown under red and blue nettings, such as shoot and fruit growth, and leaf gas exchange, and that they are probably governed by the activity of specific photoreceptors such as phytochromes and cryptochromes [11, 12, 28].

UV radiation is another component of sunlight that is affected using nets and that plays an important role in the apple fruit quality, by stimulating the synthesis of anthocyanins, the pigment responsible for the red color of the fruit [29], as well as for its action on the photo-oxidative damage that originates in the apple sunburn due to overexposure to direct sunlight [30]. It has been shown that, depending on the color and weave of the threads, the nets used to cover apple orchards reduce UV light transmission by 10–13% more than PAR light, due to the additives that are incorporated during its manufacturing to increase its durability and resistance to UV rays. Thus, for a transparent and black monofilament net that reduces PAR light by 7 and 18%, the reduction of UV light with these nets is 20 and 29%, respectively [25].

### 3.2 Temperature, relative humidity, and wind speed

Covering netting mainly reduces the global solar radiation incoming and wind speed, thus altering the heat balance in the orchard and with the consequent impact on air and canopy temperatures [31]. The density of net weaving directly affects the magnitude of temperature changes under netting. No significant changes in environmental temperature were observed under nets with low thread density and with shade levels between 20 and 25% [10, 31], while the use of 50% shade nets reduces the air temperature by up to 1–3°C, and especially with 50% black nets [32].

The effect of the nets on the relative humidity in apple orchards depends on the climatic condition, planting system, and geographical location. In more arid environments such as Australia, the use of netting increased the relative humidity by up to 15% [33], but in other arid environments such as the state of Washington, the relative humidity did not change in apple trees under different net colors [10], while in more humid environments such as Germany, the relative humidity was reduced from 3 to 5% in apple orchards grown under netting [28].

Protecting apple orchards with netting also reduces wind speed at the tree canopy level, and its effect depends on location, planting system, and density of the net weaving. In the States of Washington and Australia, a reduction of about 50% in wind speed was reported in apple trees growing under 20% shade nets [32, 34], while in Chile, the use of 23% shade net reduced the wind speed by 69% with a direct impact on the reduction of loss of sensible heat flux in apple orchards [31]. Finally, the use of colored nets at 50% shade allowed to reduce wind speed by up to 89% when compared to the condition nets [34].

## 4. Physiological tree responses

### 4.1 Leaf gas exchange

The influence of the use of the nets on apple leaf net photosynthesis rate ($A_n$) and stomatal conductance ($g_s$) has been reported with different responses, depending on the netting system, cultivar, and climatic conditions (**Table 1**). In South Africa, the use of 20% shade black net decreased $A_n$ and $g_s$ by 14 and 21% compared to the condition without netting, and attributable to morphological changes of the leaves growing under netting [35]. Similar results were observed in Germany under cloudy day conditions with the use of 12% shade green-black net in the cultivar "Fuji." In this case, the $A_n$ was 21% lower in apple leaves growing with this net [28]. Under more extreme environmental conditions with greater intensity of solar radiation and high temperatures, the use of netting has been favorable in increasing $A_n$ and $g_s$ in apple trees (**Table 1**). In the state of Washington, covering "Honeycrisp" apple trees with

| Net system/climate condition | Cultivars | Variation of $A_n$ with respect to control (%) | Variation of $g_s$ with respect to control (%) | References |
|---|---|---|---|---|
| 20% black net/Sunny condition | "Royal Gala" | (−) 14 | (−) 21 | [35] |
| 12% green-black net/ Cloudy condition | "Fuji" | (−) 21 | — | [28] |
| 22% blue net/Sunny and hot condition | "Honeycrisp" | (+) 30 | (+) 31 | [36] |
| 22% pearl-gray net/ Sunny and hot condition | "Brookfield Gala" | (+) 54 | (+) 52 | [31] |
| 22% black net/Sunny and warm condition | "Golden Delicious" | (+) 60 | (+) 80 | [37] |

**Table 1.**
*Variation in net photosynthesis rate ($A_n$) and stomatal conductance ($g_s$) in apple cultivars grown under different climate conditions and netting systems.*

a blue net at 22% shading allowed an increase of about 30% in the leaf $A_n$ and $g_s$, when the trees grew under hot conditions [36]. Similar results were obtained in Chile in "Brookfiel Gala" apple trees covered under a 22% pearl-gray net and in which $A_n$ and $g_s$ increased by 54 and 52%, respectively, when the conditions of temperature and solar radiation were widely extreme [31]. In Portugal and under Mediterranean climate conditions, the use of black netting at 22% shade in "Goden Delicious" apple allowed an increase of $A_n$ and $g_s$ by 60 and 80%, respectively [37].

Differential response in $A_n$ and $g_s$ under netting may be due to the particular light conditions that are generated under the nets and that affect the leaf function and morphology [12], as well as to the temperature conditions that prevail in the apple-growing area [36]. In this sense, the color of the net affects these responses, as has been shown in "Fuji" apple trees growing under 40% shade blue and red nets. In this research, it was shown that the $g_s$ in leaves growing under the blue net increased by 21% compared to the red net, and attributable to the effect of blue light on the stomatal opening stimulus [12]. In the same study, it was also shown that this increase in $g_s$ resulted in a similar increase in the rate of transpiration (E), but without a significant effect on An, implying a lower efficiency in the water use of the leaves that grow under the blue net. Another study indicates that the increase in $g_s$ and $A_n$ by about 50% in apple trees due to the use of 22% shade pearl-gray net also allowed for a 25% higher water use efficiency measured as $A_n/E$ [31]. These results indicate that the selection of the color and percentage of shading are very relevant to consider in apple cultivation under areas with extreme solar radiation, high temperatures, and low water availability [12, 31, 36, 37].

### 4.2 Water relations

As previously said, net application in orchards will impact the microclimate; thus, the tree's responses will also differ in terms of water use. Since there is a decrease in the incoming solar energy, the leaves will need less water for evaporation, a consequence of necessary heat loss to manage the entering energy. This amount of outcoming water flux by evaporation from the apple orchard under netting depends on the quantity of light determined by the shading percentage of the net, but also by the quality of light determined by the color of the net. Higher radiation leads to higher latent heat flow; thus, there will be higher heat dissipation [13, 38–40]. In unlimited water availability scenarios, water loss for cooling purposes will take as long as the optimum leaf temperature is maintained. However, when water is limited, stomata will close to prevent excessive water loss, thus preventing dehydration [13, 39]. Apple sap flow studies showed that trees under a 20% shading net indeed have higher xylem water transport rates, compared to trees under a 50% shading net [13]. Another study showed improved water status for apple trees under a 50% shading net [41], compared to those without net, or even to those under a 20% shading net [42]. In these cases, trees experiencing higher incoming light may have reached, or exceeded, the "threshold" at which transpiration was higher than root water uptake. Stomata closure to avoid dehydration resulted in more negative water potentials. It is likely that trees under 50% shading are growing in less stressful conditions, than those grown under lighter shading, or no shading at all. It is expected that, in a more shaded orchard, a certain amount of water may still be available in the soil [40].

These results suggest that apple trees in shaded environments are able to lower their water requirements. Since reference evapotranspiration ($Et_0$) is with doubt lower the more shade there is [43, 44], $K_c$ will be lower for apple trees in more shaded orchards (**Figure 5**) [45, 46].

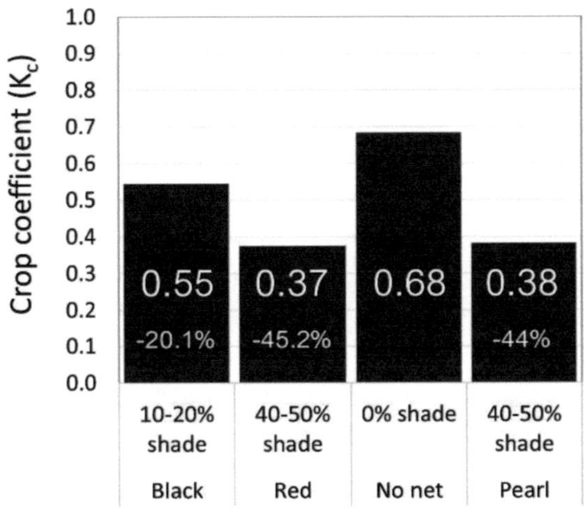


**Figure 5.**
*Estimated crop coefficient (kc) values and percentages for different light environments in a "Gala" orchard (adapted from Boini et al., 2018).*

The resulting crop evapotranspiration ($Et_c$) will decrease, compared to less shaded growing conditions. Apple trees can be more efficient in terms of water use, under shaded environments [37], even in conditions of water shortage [18]. Anyhow, increasing shade along with irrigation volumes that reflect $Et_c$ will lead to similar water potentials (**Figure 6**).

When the different wavelengths of the solar spectrum are manipulated (with photoselective nets, e.g.), water relations can be influenced according to a photomorphogenic effect. Hence, a low-red/far-red light environment will generate higher water uptake [13]. There will be a tendency to have less negative water potential [12, 13]; thus in unlimited water conditions, certain wavelengths increase tree water consumption. On the other hand, blue light was found to maintain a better water status (midday stem and leaf water potentials), compared to no shading [36], although

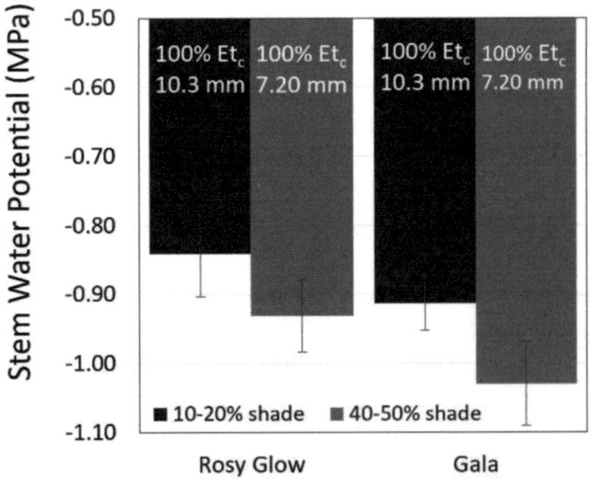


**Figure 6.**
*Daily average stem water potential for two apple cultivars in two different netted environments (unpublished data).*

there were no other colored nets as a comparison. When a photoselective net is thicker in order to increase the shading percentage (40–50% shading), the more intense certain wavebands will be. This can heavily impact the responses described. If the nets are designed to maintain a lighter shading percentage (max 20%), it is possible that the intensity of such wavebands is not strong enough to induce a visible effect, at least in the short term. Other factors will markedly dictate orchard responses, such as the pruning history and the weather Yield and fruit quality of apple cv. "Jonagold" under hail protection nets [18, 47].

### 4.3 Vegetative and reproductive growth development

Protective netting leads to higher shoot development and elongation, depending on the intensity of the external sunlight and the shading percentage of the nets. In general, lower PAR environments trigger shade-avoidance responses, to different extents [11]. As specified in the previous sub-section, shaded canopies experience low R/FR ratios, pushing shoots in search of richer PAR areas. However, there are cases that report no differences in "Fuji" shoot growth, comparing light-shading nets with uncovered controls [28], probably due to the well-known alternate bearing behavior of this cultivar.

Other vegetative growth and development features include leaf area and thickness. These characteristics will vary based on the quantity of filtered light and also on its quality. In this case, discriminating between these two properties of shading nets appears easier. Heavy shading produces smaller and thinner leaves, typical of inner canopy foliage, while the more exposed ones generate extra palisade layers, to better manage the higher incoming solar energy. In fact, extra palisade cells were found in PAR and blue-rich light environments under netting [12]. A wider leaf area is also typical of highly illuminated environments, although nets with higher PAR, or blue light, transmittance will generate bigger leaves [11, 28].

Effects of shading can occur on reproductive growth. Return bloom will be primarily influenced, as light is essential for flower bud formation; however, the responses could vary on the cultivars and the shading percentage. No differences were found for "Rosy Glow" apple, when comparing different percentage of shading (Boini, unpublished work); "Gala," "Fuji," and "Granny Smith" apple cultivars can be negatively affected by shading, with significantly fewer flower buds under more shade [28, 48, 49]. To overcome this possibility, installing reflective mulches will increase PAR distribution in the canopies, with positive results the following year [49]. Fruitlets will have to compete with vegetative shoots, for growth resources, especially in the first weeks after full bloom. In fact, this period is so crucial, and growers will not open protective netting before 3–4 weeks after full bloom. The timing depends on the location of the orchard; thus, some may see the nets open more than a month after full bloom. After such a period, there will still be competition between vegetative and reproductive organs [15]. The spur canopy needs good light interception, in order to increase the leaf area, so as to have higher photosynthates products to be sent to fruitlets. In this case, white or pearl nets with high scattering properties would improve these agronomical aspects.

### 4.4 Fruit quality traits

The location of the orchard will be the most important aspect. Different cultivating areas cannot be compared, as the tree's physiological responses will change, although the protective netting may be the same. The responses of apple fruit's final quality to shade can be conditioned by the color of the nets and by their light scattering properties. Considering black, gray, and white (pearl), the results can be different when looking at certain traits and can be considered cultivar-dependent [50].

Red color development for summer varieties can be penalized or delayed by certain protective nets, exclusion nets, for example, as daily thermal excursion may not be enough to induce anthocyanin synthesis. Nevertheless, the protective netting is beneficial against sunburn damage, as it has been demonstrated that this physiological disorder is related to high solar intensity along with high air temperatures [3]. It has been shown that the effectiveness of sunburn control through netting is related to lower photoinhibition at the level of the skin of the apple. Thus, in the cultivars "Fuji" and "Gala," the photochemical efficiency of the PS-II of the skin of the fruits that grew under nets was 3% and 12% higher in comparison with the fruits that grew exposed to full sunlight, respectively [16]. The protection of the fruits against sunburn through netting causes different responses in the composition of color pigments and antioxidants in the apple skin, which depends on the cultivar and type of net. While in "Gala" apples the use of red net favors the accumulation of anthocyanins and antioxidant capacity, and in "Fuji" apples the use of pearl net decreases the accumulation of anthocyanins and antioxidant capacity in the fruit skin [1]. Due to changes in the transmission of PAR and UV light, the antioxidant composition of anthocyanins and vitamin C is widely altered in apples that grow under the netting systems. It has been reported that the anthocyanin content was 2–6 times lower in "Fuji" apples grown under red and blue 40% shade net, in comparison with uncovered trees [51]. In the same way, it has been shown that the vitamin C content decreased by 31% in apples under the black-green net, and only by 10% under the white-red net, while in white translucent net, the vitamin C content was increased by 5% [52].

Another quality trait that is affected by the use of netting in apples is the size of the fruit, in which effect varies depending on the climatic condition, cultivar, and type of net. In Spain, the use of black 25% shade net reduced the fruit size in "Mondial Gala" apples [53], while in Brazil the use of white 18% shade net increased the fruit size in "Gala" and "Fuji" apples [48]. In Italy, the use of blue 40% shade net increased the fruit size in 'Fuji' apples, compared to the use of red 40% shade net [14]. In Chile, the use of a pearl-gray net at 22% of shading factor also increased fruit size in "Brookfield Gala" apples, attributed to the higher net assimilation of $CO_2$ found under this net and which provides greater availability of carbohydrates for fruit growth [31].

Finally, changes in fruit firmness have also been reported in apples under netting, which are closely linked to changes in light conditions. The use of a black 15% shade net increased fruit firmness in "Fuji" apples, while black 55% shade net decreased it [54]. The differential effect of the color of the net on fruit firmness in apple trees has been attributed to the role played by the light quantity and quality on changes in the size and density of cells during the fruit growth and development process [55].

## 5. Photoselective nets

In the last decades, advances in netting systems have focused on the development of photoselective materials with the capacity to transmit selectively the solar radiation to promote positive physiological responses and improve the yield and fruit quality through the addition of specific colors during the manufacture of the nets [56]. In the first studies carried out on apple trees, it was verified that the use of the

**Figure 7.**
*Detail of trial in commercial "Granny Smith" apple orchard under combined pearl-gray, blue-gray, and blue-pearl nets, in comparison with black net. Maule region, Chile.*

photo-selective blue net allowed to increase in fruit growth in comparison with the red net, and through the promotion of greater photosynthesis and partition of assimilates toward the fruit [14]. Photoselective nets also affect the sap flow and water use in apple trees; the use of the pearl net was more effective in reducing water consumption in relation to the red net, with a direct effect on decreasing the sap flow [13]. Other more recent works demonstrate the potential of using photoselective colored nets in the biological management of the codling moth in apple orchards. In this case, the parasitoid capacity of the moth larva was affected by the color of the nets; females found their host faster under red and pearl nets when compared to black nets [57]. In recent years, new photoselective net materials have been incorporated with the specific purpose of controlling sunburn on apple orchards through the reduction of light transmission in the UV and IR spectra, which are based on combined colored pearl-gray, blue-gray, and pearl-blue nets (**Figure** 7); its use in commercial apple orchards allowed reducing sun damage in fruits with an effectiveness of 49%, 45%, and 33% in the "Granny Smith," "Cripps Pink," and "Fuji" cultivars, respectively, and in relation to the use black netting [17].

## 6. Conclusion

Netting systems can increase the sustainability of apple production, limiting the use of resources, from water to chemical treatments for pest control. For this reason, this technology has been widely expanded in different apple-producing areas of the world, with different alternatives in installation structures, net design, and color that offers the possibility of differential management of the orchard microclimate and crop physiology to obtain certain benefits such as the use of water, availability of photo-assimilates, vegetative and reproductive growth control, and regulation of some fruit quality traits. The existing knowledge to date about the impacts of netting systems on plant physiology offers the possibility of exploring new applications (photoselective nets) in apple production and under different climatic conditions, but at the same time, it raises some questions from the environmental sustainability point of view. The question arises whether the actual production of the nets is going to compromise an environmentally sustainable fruit production process. The first concerns outweigh the agricultural benefits of netting on natural predators in integrated pest management [58]. A second point is related to how the netting system works well with recycling and re-uses the HDPE for constructing a series of elements, such as pumps, valves, and pipework, which is one of the best strategies [59]. Finally, increasing interest in biobased and sustainable netting systems leads to the development of polymers containing polysaccharides and raw materials. However, these biodegradable materials were still less than 1% of the produced plastics and eco-friendly additives that could extend their lifespan are still far from being produced [60].

## Author details

Richard M. Bastías[1*] and Alexandra Boini[2]

1 Department of Plant Production, University of Concepcion, Chillán, Chile

2 Department of Agricultural and Food Sciences, University of Bologna, Bologna, Italy

*Address all correspondence to: ribastias@udec.cl

## References

[1] Olivares-Soto H, Bastías RM, Calderón-Orellana A, López MD. Sunburn control by nets differentially affects the antioxidant properties of fruit peel in 'gala' and 'Fuji' apples. Horticulture, Environment, and Biotechnology. 2020;**61**:241-254. DOI: 10.1007/s13580-020-00226-w

[2] Gindaba J, Wand SJE. Comparative effects of evaporative cooling, kaolin particle film and shade net on the control of sunburn and fruit quality in apples. Hort Science. 2005;**40**:592-596. DOI: 10.21273/HORTSCI.40.3.592

[3] Kalcsits L, Asteggiano L, Schmidt T, Musacchi S, Serra S, Layne DR, et al. Shade netting reduces sunburn damage and soil moisture depletion in 'granny smith' apples. Acta Horticulturae. 2018;**1228**:85-90. DOI: 10.17660/ActaHortic.2018.1228.11

[4] Darbyshire R, McClymont L, Goodwin I. Sun damage risk of Royal Gala apple in fruit-growing districts in Australia. New Zealand Journal of Crop and Horticultural Science. 2016;**43**:222-232. DOI: 10.1080/01140671.2015.1034731

[5] Blanke MM. Perspectives of fruit research and apple orchard Management in Germany in a changing climate. Acta Horticulturae. 2008;**772**:441-446. DOI: 10.17660/ActaHortic.2008.772.75

[6] Baiamonte I, Raffo A, Nardo N, Moneta E, Peparaio M, D'Aloise A, et al. Effect of the use of anti-hail nets on codling moth (*Cydia pomonella*) and organoleptic quality of apple (cv. Braeburn) grown in alto Adige region (northern). Journal of the Science of Food and Agriculture. 2016;**96**:2025-2032. DOI: 10.1002/jsfa.7313

[7] Gonzalez L, Àvila G, Carbó J, Bonany J, Alegre S, Torres E, et al. Hail nets do not affect the efficacy of metamitron for chemical thinning of apple trees. The Journal of Horticultural Science and Biotechnology. 2020;**95**:128-135. DOI: 10.1080/14620316.2019.1631128

[8] Sauphanor B, Severac G, Maugin S, Toubon JF, Capowiez Y. Exclusion netting may alter reproduction of the codling moth (Cydia pomonella) and prevent associated fruit damage to apple orchards. Entomologia Experimentalis et Applicata. 2012;**145**:134-142. DOI: 10.1111/j.1570-7458.2012.01320.x

[9] Chouinard G, Veilleux J, Pelletier F, Larose M, Philion V, Joubert V, et al. Impact of exclusion netting row covers on 'Honeycrisp' apple trees grown under northeastern north American conditions: Effects on photosynthesis and fruit quality. Insects. 2019;**10**:214. DOI: 10.3390/insects10070214

[10] Kalcsits L, Musacchi S, Layne DR, Schmidt T, Mupambi G, Serra S, et al. Above and below-ground environmental changes associated with the use of photoselective protective netting to reduce sunburn in apple. Agricultural and Forest Meteorology. 2017;**237-238**:9-17. DOI: 10.1016/j.agrformet.2017.01.016

[11] Bastías RM, Corelli-Grappadelli L. Light quality management in fruit orchards: Physiological and technological aspects. Chilean Journal of Agriculture Research. 2012;**72**:574-582. DOI: 10.4067/S0718-58392012000400018

[12] Bastías RM, Losciale P, Chieco C, Corelli-Grappadelli L. Red and blue netting alters leaf morphological and physiological characteristics in apple

trees. Plants. 2021;**10**:127. DOI: 10.3390/plants10010127

[13] Boini A, Bresilla K, Perulli GD, Manfrini L, Corelli Grappadelli L, Morandi B. Photoselective nets impact apple sap flow and fruit growth. Agricultural Water Management. 2019;**226**:105738. DOI: 10.1016/j.agwat.2019.105738

[14] Bastías RM, Manfrini L, Corelli-Grappadelli L. Exploring the potential use of photoselective nets for fruit growth regulation in apple. Chilean Journal of Agriculture Research. 2012;**72**:224-231. DOI: 10.4067/S0718-58392012000200010

[15] Boini A, Casadio N, Bresilla K, Perulli GD, Manfrini L, Corelli Grappadelli L, et al. Early apple fruit development under photoselective nets. Scientia Horticulturae. 2022;**292**:110619. DOI: 10.1016/j.scienta.2021.110619

[16] Olivares-Soto H, Bastías RM. Photosynthetic efficiency of apples under protected shade nets. Chilean Journal of Agriculture Research. 2018;**78**:126-138. DOI: 10.4067/S0718-58392018000100126

[17] Bastías RM, Moya E, Quezada C, Leyton MJ, Castro da Costa D, Pinto A. Manejo de huertos de manzanos bajo mallas para el control de golpe de sol [Apple orchard management under nets for sunburn control]. Santiago: Fundación para la Innovación Agraria; 2018 106 p.

[18] Boini A, Bortolotti G, Perulli GD, Venturi M, Bonora A, Manfrini L, et al. Late ripening apple production benefits from high shading and water limitation under exclusion netting. Horticulturae. 2022;**8**:884. DOI: 10.3390/horticulturae8100884

[19] Manja K, Aoun M. The use of nets for tree fruit crops and their impact on the production: A review. Scientia Horticulturae. 2019;**246**:110-122. DOI: 10.1016/j.scienta.2018.10.050

[20] Castellano S, Scarascia-Mugnozza G, Russo G, Briassoulis D, Mistriotis A, Hemming S, et al. Design and use criteria of netting systems for agricultural production in Italy. Journal of Agricultural Engineering. 2008;**3**:31-42. DOI: 10.4081/jae.2008.3.31

[21] Mistriotis A, Briassoulis D. Windloads on net covered structures. Acta Horticulturae. 2008;**801**:963-970. DOI: 10.17660/ActaHortic.2008.801.114

[22] Mupambi G, Layne DR, Kalcsits L, Musacchi S, Serra S, Schmidt T, Hanrahan I. Use of protective netting in Washington State apple production [Internet]. 2019. Available from: https://tfrec.cahnrs.wsu.edu/kalcsits/wp-content/uploads/sites/7/2019/10/Mupambi-etal-2019-TB60E-Netting.pdf. [Accessed: October 16, 2022].

[23] Briassoulis D, Mistriotis A, Eleftherakis D. Mechanical behaviour and properties of agricultural nets. Part II: Analysis of the performance of the main categories of agricultural nets. Polymer Testing. 2007;**26**:970-984. DOI: 10.1016/j.polymertesting.2007.06.010

[24] Corelli-Grappadelli L. Light relations. In: Ferree DC, Warrington IJ, editors. Apples, Production, Botany, and Uses. Wallingford, UK: CAB International; 2003. pp. 195-216

[25] Blanke M. Farbige Hagelnetze: Ihre Netzstruktur sowie Licht- und UV-Durchlassigkeit bestimmen die Ausfarbung der Apfelfruchte. Ewerbs-Obstbau. 2007;**2007**(49):127-139. DOI: 10.1007/s10341-007-0048-6

[26] Umanzor C, Bastías RM, Wilckens R. Quezada C influence of

using of pearl and red nets on micro-climate conditions, fruit sun damage and fruit color development in 'gala' and 'Fuji' apple orchards. Revista Brasileira de Fruticultura. 2017;**39**:930-943. DOI: 10.1590/0100-29452017930

[27] Shahak Y, Ratner K, Giller Y, Zur N, Or E, Gussakovsky E, et al. Improving solar energy utilization, productivity and fruit quality in orchards and vineyards by photoselective netting. Acta Horticulture. 2008;**772**:65-72. DOI: 10.17660/ActaHortic.2008772.7

[28] Solomakhin A, Blanke M. Coloured hailnets alter light transmission, spectra and phytochrome, as well as vegetative growth, leaf chlorophyll and photosynthesis and reduce flower induction of apple. Plant Growth Regulation. 2008;**56**:211-218. DOI: 10.1007/s10725-008-9302-7

[29] Solovchenko A, Schmitz-Eiberger M. Significance of skin flavonoids for UV-B-protection in apple fruits. Journal of Experimental Botany. 2003;**54**:1977-1984. DOI: 10.1093/jxb/erg199

[30] Felicetti DA, Schrader LE. Photooxidative sunburn of apples: Characterization of third type of apple sunburn. International Journal of Fruit Science. 2008;**8**(3):160-172. DOI: 10.1080/15538360802526472

[31] Bastías RM, Matamala MF, Leyton MJ, Soto G, González E, Urra I, et al. An energy balance approach for precision physiology-based management of apple orchards under netting. Acta Horticulturae. 2021;**1314**:313-319. DOI: 10.17660/ActaHortic.2021.1314.39

[32] Mupambi G, Anthony BM, Layne DR, Musacchi S, Serra S, Schmidt T, et al. The influence of protective netting on tree physiology and fruit quality of apple: A review. Scientia Horticulturae. 2018;**236**:60-72. DOI: 10.1016/j.scienta.2018.03.014

[33] Middleton S, McWaters A. Hail netting of apple orchards—Australian experience. Compact Fruit Tree. 2002;**35**:51-55

[34] Arthurs SP, Stamps RH, Giglia FF. Environmental modification inside photoselective shadehouses. Hort Science. 2013;**48**:975-979. DOI: 10.21273/HORTSCI.48.8.975

[35] Gindaba J, Wand SJ. Climate-ameliorating measures influence photosynthetic gas exchange of apple leaves: Research article. The Annals of Applied Biology. 2007;**150**:75-80. DOI: 10.1111/j.1744-7348.2006.00110.x

[36] Mupambi G, Musacchi S, Serra S, Kalcsits LA, Layne DR, Schmidt T. Protective netting improves leaf-level photosynthetic light use efficiency in 'Honeycrisp' apple under heat stress. Hortscience. 2018;**53**:1416-1422. DOI: 10.21273/HORTSCI13096-18

[37] Brito C, Rodrigues MÂ, Pinto L, Gonçalves A, Silva E, Martins S, et al. Grey and black anti-hail nets ameliorated apple (Malus × domestica Borkh. cv. Golden delicious) physiology under Mediterranean climate. Plants. 2021;**10**:2578. DOI: 10.3390/plants10122578

[38] Green S, McNaughton K, Wünsche JN, Clothier B. Modeling light interception and transpiration of apple tree canopies. Agronomy Journal. 2003;**95**:1380-1387. DOI: 10.2134/agronj2003.1380

[39] Nicolás E, Torrencillas A, Dell'Amico J, Alarcón JJ. Sap flow, gas exchange, and hydraulic conductance of young apricot trees growing under a shading net and different water

supplies. Journal of Plant Physiology. 2004;**162**:439-447. DOI: 10.1016/j.jplph.2004.05.014

[40] McCaskill MR, McClymont L, Goodwin I, Green S, Partington DL. How hail netting reduces apple fruit surface temperature: A microclimate and modelling study. Agricultural and Forest Meteorology. 2016;**226-227**:148-160. DOI: 10.1016/j.agrformet.2016.05.017

[41] Lopez G, Boini A, Manfrini L, Torres-Ruiz JM, Pierpaoli E, Zibordi M, et al. Effect of shading and water stress on light interception, physiology and yield of apple trees. Agricultural Water Management. 2018;**210**:140-148. DOI: 10.1016/j.agwat.2018.08.015

[42] Boini A, Manfrini L, Morandi B, Corelli Grappadelli L, Predieri S, Daniele GM, et al. High levels of shading as a sustainable application for mitigating. Agronomy. 2021;**11**(3):422. DOI: 10.3390/agronomy11030422

[43] Katerji N, Rana G. Modelling evapotranspiration of six irrigated crops under Mediterranean climate conditions. Agricultural and Forest Meteorology. 2006;**138**:142-155. DOI: 10.1016/j.agrformet.2006.04.006

[44] Lecina S, Martínez-Cob A, Pérez PJ, Villalobos FJ, Baselga JJ. Fixed versus variable bulk canopy resistance for reference evapotranspiration estimation using the penman-Monteith equation under semiarid conditions. Agricultural Water Management. 2003;**60**:181-198. DOI: 10.1016/S0378-3774(02)00174-9

[45] Auzmendi I, Mata M, Lopez G, Girona J, Marsal J. Intercepted radiation by apple canopy can be used as a basis for irrigation scheduling. Agricultural Water Management. 2011;**98**(5):886-892. DOI: 10.1016/j.agwat.2011.01.001

[46] Boini A, Lopez G, Morandi B, Manfrini L, Corelli-Grappadelli L. Testing the effect of different light environments and water shortage on apple physiological parameters and yield. Acta Horticulturae. 2018;**1228**:397-403. DOI: 10.17660/ActaHortic.2018.1228.59

[47] Stampar F, Veberic R, Zadravec P, Hudina M, Usenik V, Solar A, et al. Yield and fruit quality of apples cv. 'Jonagold' under hail protection nets/Ertrag und fruchtqualität der apfelsorte 'Jonagold' unter hagelschutznetzen. Die. Gartenbauwissenschaft. 2002;**67**:205-210

[48] Do Amarante CVT, Steffens CA, Argenta LC. Yield and fruit quality of 'gala' and 'Fuji' apple trees protected by white anti-hail net. Scientia Horticulturae. 2011;**129**:79-85. DOI: 10.1016/j.scienta.2011.03.010

[49] Mupambi G, Schmidt T, Valverdi N, Camargo-Alvarez H, Reid M, Kalcsits L. Reflective groundcover improves return bloom and yield in 'granny smith' apple under protective netting. Acta Horticulturae. 2022;**1337**:409-416. DOI: 10.17660/ActaHortic.2022.1337.56

[50] Bosančić B, Mićić N, Blanke M, Pecina M. A main effects meta principal components analysis of netting effects on fruit: Using apple as a model crop. Plant Growth Regulation. 2018;**86**:455-464. DOI: 10.1007/s10725-018-0443-z

[51] Bastías RM, Ruíz K, Manfrini L, Pierpaoli E, Zibordi M, Morandi B, et al. Effects of Photoselective nets on phenolic composition in apple fruits. Acta Horticulture. 2012;**939**:77-84. DOI: 10.17660/ActaHortic.2012.939.9

[52] Solomakhin A, Blanke M. Can coloured hailnets improve taste (sugar, sugar: Acid ratio), consumer appeal (colouration) and nutritional value (anthocyanin, vitamin C) of apple

fruit? Food Science and Technology. 2010;**43**:1277-1284. DOI: 10.1016/j.lwt.2010.02.020

[53] Iglesias I, Alegre S. The effect of anti-hail nets on fruit protection, radiation, temperature, quality and profitability of 'Mondial gala' apples. Journal of Applied Horticultre. 2006;**8**:91-100

[54] Dussi MC, Giardina G, Sosa D, González-Junyent R, Zecca A, Reeb P. Shade nets effect on canopy light distribution and quality of fruit and spur leaf on apple cv. Fuji. Spanish Journal of Agricultural Research. 2005;**2005**(3):253-260

[55] Corollaro ML, Manfrini L, Endrizzi I, Aprea E, Dematte' ML, Charles MC, et al. The effect of two orchard light management practices on the sensory quality of apple: Fruit thinning by shading or photo-selective nets. Journal of Horticulture Science and Biotechnology. 2015;**90**(1):99-107. DOI: 10.1080/14620316.2015.11513159

[56] Shahak Y, Gussakovsky EE, Cohen Y, Lurie S, Stern R, Kfir S, et al. ColorNets: A new approach for light manipulation in fruit trees. Acta Horticulturae. 2004;**636**:609-616. DOI: 10.17660/ActaHortic.2004.636.76

[57] Yáñez-Díaz MJ, Rodríguez M, Musleh S, Devotto L, Silva G, Lucas E. Photo-selective nets and Pest control: Searching behavior of the codling moth parasitoid Mastrus ridens (hymenoptera: Ichneumonidae) under varying light quantity and quality conditions. Insects. 2021;**12**:582. DOI: 10.3390/insects12070582

[58] Maraveas C. The sustainability of plastic nets in agriculture. Sustainability. 2020;**12**:3625. DOI: 10.3390/su12093625

[59] Pietrelli L. Polypropylene recovery and recycling from mussel nets. Polymers. 2022;**14**:3469. DOI: 10.3390/polym14173469

[60] Mukherjee A, Knoch S, Chouinard G, Tavares JR, Dumont MJ. Use of bio-based polymers in agricultural exclusion nets: A perspective. Biosystems Engineering. 2019;**180**:121-145. DOI: 10.1016/j.biosystemseng.2019.01.017

Chapter 8

# Rootstocks and Varieties of Fruits, Berry Crops, and Grapes, Used for Intensive Gardening in Kazakhstan

*Yuliya M. Yefremova, Marina V. Urazayeva and Saule Sh. Kazybaeva*

## Abstract

The results of the study of clonal rootstocks of apple trees of various ecological and geographical origins, varieties of fruit, berry crops, and grapes of local selection are presented. According to the efficiency of reproduction in the mother liquor and fields of nursery formation, as well as the short stature of trees of grafted varieties in the garden, precocity and productivity, clonal rootstocks of the apple tree B7-35, Arm18, 62-396, and B16-20 were distinguished. Their important role in unlocking the potential of ancient apple varieties in the south and southeast of Kazakhstan. These rootstocks and varieties are recommended for propagation and creation of highly productive orchards. The Ministry of Agriculture of the Republic of Kazakhstan provided funding within the framework of the budget program 267 "Improving the availability of knowledge and scientific research" under subprogram 101 "Program-targeted financing of scientific research and activities" – Creation of varieties and hybrids of fruit and berry, nut crops and grapes based on the achievements of bio and IT technologies, 2021–2023yy. No. BR10765032.

**Keywords:** clonal rootstocks, nursery, seedlings, varieties, intensive gardening, selection

## 1. Introduction

In recent years, the economy of Kazakhstan has been steadily outpacing the average world growth rates, ensuring the progressive socioeconomic development of the country. In 2019, GDP growth amounted to 4.5%, accelerating the momentum gained in 2017–2018. More than 85% of economic growth in 2019 was provided by the non-primary sector of the economy. These are construction, manufacturing, and services in general. That's where the growth prospects for the village. Therefore, our task is to at least increase productivity in the agro-industrial complex by 2024. Only agro-industrial diversification, that is, a sharp increase in the processing of agricultural raw materials, new equipment, new technologies, and approaches in agriculture, can solve this difficult task. It is necessary to use world experience, to quickly introduce it into our agriculture [1].

At present, in Kazakhstan, the production of fruit and berry products, as well as early vegetables during the off-season, is not sufficient. At the same time, there is a great potential for increasing the production of these products, given the favorable natural and climatic conditions of the southern regions of the republic for the cultivation of fruit crops and grapes. There is also a large market for fruit products in the border regions of the Russian Federation. One of the effective ways to increase the production of these products is the introduction of advanced technologies, in particular, intensive orchards with a high planting density, accelerated entry into commercial fruiting, and high yield potential. In order to implement the Program for the Development of the Agro-Industrial Complex of the Republic of Kazakhstan for 2013–2020 "Agribusiness-2020," a Master Plan "Fruit and Vegetable Growing" was developed [2].

The main requirements of industrial horticulture are to obtain early-growing, high-yielding and low-growing fruit trees. These main requirements of intensification are achieved by introducing new varieties and rootstocks into the assortment, using various agro-technical methods, and using completely new types of planting structures [3, 4].

In the practice of managing the genetic diversity of planting material in Kazakhstan, a course has been taken to create gardens on low-growing vegetatively propagated (clonal) rootstocks, as well as to obtain highly adaptive varieties of horticultural crops by breeding [5].

In this regard, F&VRI has obtained a unique collection of rootstocks and varieties from various fruit growing regions of the world, as well as varieties of local selection, which has replenished the agro-diversity of fruit crops in Kazakhstan. Long-term studies have identified a group of rootstocks with positive economic and biological characteristics, in particular, good adaptability to local soil and climatic conditions of growth, contributing to high productivity of trees with good fruit quality [6].

At present, the isolated rootstocks of apple trees and varieties of horticultural crops form the basis for the development of a system for conducting intensive horticulture in Kazakhstan. The use of their biological potential will significantly increase the profitability of orchards in newly created peasant and farm enterprises and support their economy [7].

## 2. Materials and methods

The objects of research in the mother liquor, fruit nursery, and orchards were new for Kazakhstan clonal apple rootstocks and varieties of local selection of horticultural crops. The studies were carried out in the mother liquor of vegetatively propagated rootstocks, in the fields of nursery formation and in gardens of Almaty, Zhambyl, and South Kazakhstan regions of the Republic of Kazakhstan. Over the years of research, 150 clonal rootstocks of apple trees were studied, including 14 forms of apple trees of the M series, No. 34–30, No. 34–38, (England, East Malling Research Institute of Horticulture), 12 types of the MM series (England, East Malling Research Institute and D. Innes Research Institute of Horticulture), 10 rootstocks bred by V.I. Budagovsky, 4 rootstocks of the North Caucasian Zonal Research Institute of Horticulture and Viticulture, 52 forms of the B series (Dagestan, Buynaksk Experimental Station), 18 types of the Arm and LA series (Armenia, NIIVV and C), etc. The controls are generally recognized in world practice dwarf rootstocks M9, srednerosly MM106, in the gardens of the seed rootstock Nedzvetsky's apple tree. Introduced and domestic varieties and hybrids of grapes. The genetic fund of grapes in Kazakhstan is more than 500 variety samples, where varieties are collected from almost all viticulture regions of the world, of which 28 varieties are selected by the Research Institute of Fruit Growing and Viticulture [8].

The soils of the experimental plot are dark chestnut, low humus, the average content of humus in the arable horizon is 2.3% with a deep occurrence of pebbles. The soils of the experimental plots before the laying of the experiments had an average supply of hydrolyzable nitrogen, mobile phosphorus, exchangeable potassium, zinc, low boron, and high manganese [9].

According to the mechanical composition, the soils are medium loamy. Soil-forming rocks are deposits represented by loess-like loamy clays, which are characterized by an increased content of carbonates. The soil reaction is alkaline, pH – 7.8.

The studies were carried out according to the methodological recommendations of the Kazakh Research Institute of Fruit Growing and Viticulture [10], All-Russian Research Institute of Horticulture. I.V. Michurin [11], Uman Agricultural Institute [12], scientific institutions of the Baltic Republics and Belarus [13].

## 3. Results and discussion

The manufacturability of the rootstock in the mother liquor of vegetatively propagated sub-roots was determined by such indicators as the shoot-producing ability of mother bushes, the absence of shoots with lateral branches, the number of rooted shoots, the degree of rooting, and the output of standard layering.

For many years, the introduction and study of seed and clonal rootstocks of apple, pear and stone fruit species have been carried out in the nursery and orchards. Experimental gardens were laid in Almaty, Zhambyl, South Kazakhstan regions. Combinations of rootstocks with apple tree varieties Aport, Zarya Alatau, Golden Delicious, Milton, Jonathan were studied. Semi-dwarf – B 16–20.

**Arm 18.** Bred at the Armenian Research Institute of Viticulture, Winemaking and Fruit Growing (Yerevan) by breeder L.A. Apoyan.

The uterine bush is undersized, bushy in shape. The height of the uterine bush is 40 cm. The rooting of the layers is excellent – 4–5 points, the diameter of the conditional root collar is 8 mm. Layers are characterized by a strongly developed fibrous root system with good regeneration after planting. The output of standard cuttings in the mother liquor of vegetatively propagated rootstocks is 250–350 thousand pieces/ha, or an average of 300 thousand pieces/ha, which is significantly higher than the analogue of M9 and other clonal apple rootstocks. One of the best clonal apple rootstocks, perfectly propagated by woody cuttings in the open field.

Apple-tree varieties Aport, Golden Delicious, Jonathan, Zarya Alatau, Milton, Saltanat budded on Arm 18 give a high yield of standard 1-year-old seedlings, an average of 52 thousand pcs/ha for varieties or 99% of the total number of dug seedlings, of which 77% first commercial grade.

The trees in the garden are stunted and smaller in size than M9, early-bearing, they bear fruit 4–5 years after planting and subsequently are characterized by high yields. The average yield of various varieties is 180–260 q/ha. Trees are well fixed in the soil. The rootstock is more drought-resistant than M9.

The stock is recommended for use in Almaty, Zhambyl, and South Kazakhstan regions.

Bush of dwarf rootstock Arm 18

Rooting of shoots in the bush of the rootstock Arm18

**B7-35.** Received at the Buinaksk Experimental Horticulture Station D.N. Krylov, R.G. Tsabolov (Dagestan, Buynaksk).

The uterine bush is medium tall, bushy-pyramidal in shape. In the mother liquor, the height of the bush is 62 cm. Rooting of cuttings takes an average of 8 years, 3–10 years after planting – 4.4 points. The uterine bushes are resistant to falling out. From the moment of planting and during 12 years of operation of the mother liquor, 97% of the bushes were preserved, in M9 – 86%. The output of standard cuttings is high – 237 thousand pieces/ha, which is 83% of the total number of shoots.

In the nursery, this stock has even, smooth trunks with elastic bark, which makes budding easier. The output of seedlings in the nursery is 47,000 pieces/ha of 1-year-olds.

Apple cultivars in the orchard are characterized by moderate growth or level with trees on M9.

Apple varieties begin to bear fruit 3–4 years after planting. The average yield of 14 18-year-old trees is 130 (Aport) and 230 q/ha (Zarya Alatau, Jonathan). Trees of grafted varieties on this stock, unlike M9, are very firmly fixed in the soil due to the presence of skeletal roots. A valuable feature of B7-35 is its high resistance to drought, which is of great importance for a dwarf rootstock under irrigated fruit growing.

For high economic and biological indicators, the dwarf apple rootstock B7-35 is recommended for use in Almaty, Zhambyl, and South Kazakhstan regions.

Bush of dwarf rootstock B7-35

Rooting shoots in a rootstock bush B7-35

**62-396.** Selections of the Department of Fruit Growing of the Michurinsk State Agrarian University, selected from the hybrid fund of V.I. Budagovsky and others (Russia, Michurinsk).

The dwarf rootstock of the apple tree has an anthocyanin coloration of leaves, shoots, bark, wood, and roots inherited from the Nedzvetsky apple tree – a local species of Kazakhstan.

In the mother liquor of clonal rootstocks of apple tree 62–396, it is characterized by excellent rooting of shoots – 4.7 points and a high yield of standard cuttings – 260,000 pieces/ha, which is four times more than M9. Thanks to good rooting of the shoots and a powerful fibrous root system, they take root well and grow intensively in the nursery, which makes it possible to obtain a high yield of standard seedlings.

The height and size of trees on this rootstock are 14–28% lower than on M9, depending on the variety. They begin to bear fruit 4–5 years after planting. The average yield of various varieties is 150–220 q/ha. Under-howl is very drought-resistant and winter-hardy, having inherited valuable traits from Nedzwiecki's apple tree. The rootstock can be successfully used in the northern regions of Kazakhstan.

Since 1998, the stock has been included in the State Register of the Republic of Kazakhstan and approved for use in Almaty, Zhambyl, South Kazakhstan regions.

**Dwarf rootstock bush 62-396**

**Rooting shoots in a rootstock bush 62-396**

Employees of the nursery laboratory, as a result of crossing clonal rootstocks with the Nedzvetsky apple tree, received a series (seven forms) of new low-growing vegetatively propagated rootstocks called "Zhetysu." The rootstock Zhetysu 5 turned out to be promising for production.

**Zhetysu 5.** Selections of LLP "KazNII of fruit growing and viticulture."

The shoots of the rootstock Zhetysu 5 are characterized by very high rooting in the mother liquor – 4.4 points. The output of standard cuttings is very high – 321,000 pieces/ha. Thanks to a powerful root system, they take root well and grow intensively in the nursery, which makes it possible to obtain a high yield of standard seedlings.

The height of 14-year-old Golden Delicious apple trees under irrigated conditions in the foothill zone is 3.2 m, which is 20% higher than on M9, the crown width is 2.3 m, the projection area is 4.1 $m^2$, and the volume is 6.0 $m^3$.

The apple-tree variety Golden Delicious on the rootstock Zhetysu 5 begins to bear fruit 5 years after planting. The average yield of Golden Delicious trees is 171.5 q/ha. Rootstock is very drought-tolerant. It is a promising semi-dwarf apple rootstock for the conditions of the south-east of Kazakhstan.

**Bush semi-dwarf rootstock Zhetysu 5**

**Rooting shoots in the bush of the rootstock Zhetysu 5**

Variety Aport gave the highest total yield on a dwarf rootstock B 7-35 (1758 q/ha). As seed rootstocks for the variety Aport, 22 varieties of apple trees were studied, of which, according to positive economic and biological indicators, seedlings of the varieties Pestrushka and Eko-numberat Extermayer stood out.

For pear varieties Forest Beauty and Talgar Beauty, low-growing seed rootstocks were selected. It has been established that among 23 pear seed rootstocks, the East Asian group of the Bretschneider species is characterized by an excellent root system in the nursery, a high yield of seedlings. Variety Forest beauty on rootstocks of Chinese origin Xiang li, Bai li, Xiao he bai li, Zi li, in comparison with the forest pear, differs in moderate growth, compact crown (height of 6 summer trees – 1.9–2.8 m, on forest pear – 3.1 m).

Out of 23 forms of pear clonal rootstocks, a high yield of standard offspring in the mother liquor was obtained from quince Arm 21, No. 1, K13, Sido.

A large collection of seed and clonal rootstocks for stone fruits has been studied. There are 36 types and forms in total. The collection of plum seed rootstocks includes cherry plums of the North Caucasus, plums of the Mountain Pa-mir (Balzhuanskaya, Darvazskaya), hybrids of cherry plum, and Aflatunia, felt cherry from the Far East. The most effective rootstock for plum varieties Victoria, Stanley, was felt cherry. The average yield for the fruiting years of the Stanley variety was 240 c/ha, of the Victoria variety –120 c/ha.

For cherry from the studied forms of cerapadus (VP-1, 28,888, 30,020, 31,409), the rootstock VP-1 was isolated. Cherry cultivars Lyubskaya and Komsomolskaya proved to be more productive on this rootstock than on Maga-lebka cherry.

New rootstocks, identified as a result of many years of work at the Kazakh Research Institute of Fruit Growing and Viticulture, made it possible to reevaluate the potential of a number of ancient and local varieties of the south and south-east of Kazakhstan, such as Aport, Zarya Alatau, Salta-nat, and others. These varieties, having excellent taste and appearance, as a rule, did not differ in consistently high yields over the years. On new rootstocks, they gave a significantly greater economic effect and therefore should not be discounted when determining the assortment for laying commercial gardens and growing planting material. So, for example, the unique variety Aport on the dwarf rootstock B7-35 exceeded by 35–45%, in terms of productivity, trees on M 9. 9 times higher than on the widespread rootstock MM106.

In 2015, a mother plant was planted with clonal rootstocks of stone fruit crops bred by Eremina G.V. (11 forms). According to economic and biological characteristics in the mother liquor, the forms VSV-1, Druzhba, Evrika 99 can be attributed to the group of promising clonal rootstocks of stone fruit crops [14].

At present, the ampelographic collection is located in two different natural and climatic zones: in the south near the city of Shymkent and in the southeast, near the city of Almaty.

Over the years of the existence of the ampelographic collection, the grape gene pool has been drawn from 22 countries, almost from all regions of viticulture, and amounts to more than 500 varieties, of which 28 varieties are bred by the Kazakh Research Institute of Fruit Growing and Viticulture [15].

In the ampelographic collection, a group of oriental varieties, Taifi pink, Nimrang, Boyan shirey, Khindogny, Khu-saine, Karaburnu, etc., is widely represented. Moldova, Queen of Vineyards, and many others. Western European group – varieties Aligote, Pinot franc, Riesling, Cabernet franc, etc. American group – varieties Lydia, Isabella, Lyatis.

Also in the ampelographic collection, there are native varieties – Kuldzhinsky, Uigursky white.

Breeding is inextricably linked with variety study, since without knowledge of grape varieties and their characteristics, it is impossible to correctly select the source material for breeding new varieties. The collection has a hybrid nursery, which contains more than 3500 hybrid forms, from which a rigorous selection is carried out annually according to a set of positive indicators [16].

All vineyards in Kazakhstan are located in the zone of sheltered viticulture; therefore, it is important to select varieties that are characterized by increased winter hardiness suitable for growing grapes in a subclinical and even non-covering culture. On this basis, nine varieties were selected, as well as breeders of Kazakhstan created a group of varieties with increased winter hardiness (Samal, Almaly, Iliysky, Bereke), which allow the cultivation of grapes in a subclinical culture.

In Kazakhstan, in recent years, epiphytoties of mildew and oidium have become more and more frequent. As a result of the study and analysis, 12 varieties were selected from the collection, which most fully meet modern requirements (Moldova, December, Lyana, Rusmol, Citron Magaracha, etc.). All of these cultivars are used in crossbreeding as parental forms to create complex disease-resistant cultivars [17].

Kazakhstan is in the phylloxera free zone. Until now, all vineyards are root-owned, so at present, the import of foreign collection material is possible only through in vitro, and this is associated with additional difficulties. In this regard, selection is underway to create local varieties of ultra-early and early ripening, as well as varieties with high winter hardiness and disease resistance, high yields. New varieties and hybrid forms (Kara-Koz,

Almaty, Kyzyl Tan, Aigul, Kishmish Alma-Ata, Muscat Kazakhstani, etc.) in the south and south-east of Kazakhstan are far ahead of standard varieties in terms of maturation. Many of them have already been zoned and are undergoing state testing [18].

**Table grape varieties**

**Grape variety Alma-Ata early**

**Year of inclusion of the variety in the State Register:** 1974

**Authors:** Ponamarchuk V.P., Bogdanova V.S.

**Originator:** Kazakh Research Institute of Horticulture LLP

**Origin:** Madeleine Angevin.

**General characteristics:** Very early table variety.

**Approbation signs:** Bisexual flower. The berries are medium, round, light green, golden yellow on the sunny side. The skin of the berries is thin, well-eaten, the flesh is juicy, fleshy. The taste of berries is pleasant with a nutmeg aroma. The bushes are medium tall, the ripening of the shoots is good.

**Productivity:** Productivity is 120–130c/ha, differs in good sa-horonaccumulation.

**Resistance to diseases and climatic conditions:** Winter hardiness is relatively high. Disease resistance is good.

**Recommendations:** Zhambyl region (**Table 1**).

**Grape variety Alma-Ata**

| Rootstocks | From a bush, pcs. | thousand units/ha | Standard of the total number of shoots,% | First commercial grade from standard layers, % |
|---|---|---|---|---|
| M9 (St) | 2.1 | 65 | 87 | 62 |
| MM106 (St) | 3.4 | 106 | 74 | 52 |
| Arm18 | 11.2 | 353 | 88 | 52 |
| 62–396 | 8.3 | 259 | 85 | 61 |
| B7–35 | 7.6 | 237 | 83 | 48 |
| SPS -7 | 4.1 | 128 | 53 | 79 |
| B16–20 | 6.2 | 194 | 91 | 52 |

**Table 1.**
*Yield of standard apple cuttings depending on rootstocks (average over 8 years).*

**Year of inclusion of the variety in the State Register:** 2004
**Authors:** Ponamarchuk V.P., Tekhneryadnova R.T.
**Originator:** Kazakh Research Institute of Horticulture LLP
**Origin:** It was bred by the method of intraspecific hybridization from crossing varieties Druzhba X Rizamat and subsequent individual selection.
**General characteristics:** Table variety of medium-late ripening period. It is offered as a table variety for fresh consumption (**Table 2**).
**Approbation characteristics:** A cluster of large size, conical shape, medium-dense, stem length 3 cm. The average weight of the clusters is 410 g. The berry is large, oval shape. The color of the berries is black. The skin is rough. The consistency of the pulp is fleshy-juicy, the juice is not colored, the taste is pleasant, harmoniously sweet. Seeds – 2–3 pieces, seed size is medium, pear-shaped, light brown in color. The sugar content in berry juice is 17%. Acidity −6.2 g/l. Tasting score 4.8 points.
**Productivity:** Productivity from a bush is 6.8 kg, productivity from 1 hectare is 165.9 q/ha.
Resistance to diseases and climatic conditions: Covering culture. Winter hardiness is average. Disease resistance is average.
**Recommendations:** Zhambyl region.

| Rootstocks | Output of one-year-olds | | |
|---|---|---|---|
| | from 1 ha, thousand pieces | Of the total number of seedlings, % | First grade from standard |
| M9 (St) | 54.0 | 97 | 77 |
| MM106 (St) | 65.4 | 98 | 85 |
| Arm 18 | 72.2 | 99 | 77 |
| 62–396 | 68.6 | 98 | 83 |
| B7–35 | 64.6 | 99 | 82 |
| B16–20 | 74.0 | 99 | 87 |

**Table 2.**
*Influence of rootstocks on the output of standard annual apple seedlings in the nursery.*

**Grape variety Muscat Kazakhstani**

**Year of inclusion of the variety in the State Register:** 2011

**The authors:** Ponomarchuk V.P., Tekhneryadnova R.T., Karycheva L.A., Smurygin A.S.

**Originator:** Kazakh Research Institute of Horticulture LLP

**Origin:** Madeleine Ange. x Muscat of Alexandria. + Muscat Uzbekistan.

**General characteristics:** Table variety of medium ripening period.

**Approbation signs:** Bisexual flower. The berries are medium and large, white, oval, the flesh is fleshy, with a nutmeg flavor. Bushes of medium strength, shoots ripen satisfactorily.

**Productivity:** Productivity average (140–160c/hectare).

**Flour-grinding and baking qualities:** Clusters are medium and large (180–260 g), medium dense.

**Disease and weather resistance:** Disease resistance is moderate.

**Recommendations:** Almaty region.

**Grape variety Bereke**

**Year of inclusion of the variety in the State Register:** 2020

**Authors:** Ponomarchuk V.P., Tekhneryadnova R.T., Karycheva L.A., Beketaeva L. I., Kazybaeva S.Zh.

**Originator:** Kazakh Research Institute of Horticulture LLP

**Origin:** Northern x Ili.

**General characteristics:** Variety of early-medium ripening, technical direction of use.

**Approbation signs:** The berry is small, black, rounded. The bunch is medium, dense. Resistant to oidium and frost. The bushes are very tall, the ripening of the shoots is good. Recommended for making intensely colored dry and dessert wines.

**Productivity:** Productivity is 90–104 c/hectare.

**Resistance to diseases and climatic conditions:** The variety is relatively resistant to mildew, oidium, and low temperatures.

**Recommendations:** Turkestan region.

**Apple tree**

**Apple variety Ainur**

**Year of inclusion of the variety in the State Register:** 2011

**Authors:** Vinovets A.D., Ostarkova L.V.

**Originator:** Kazakh Research Institute of Horticulture LLP

**Origin:** Created by the method of intraspecific hybridization of individual selection from crossing Aport x Golden Delicious.

**General characteristics:** Variety of autumn–winter ripening period. Winter hardy. Disease resistant.

**Approbation signs:** The tree is medium-sized, the crown is round, sprawling, medium thickened. Fruits of medium size 170–200 g, round-conical, golden yellow, with a slight blush, sweet and sour taste, with a strong aroma, creamy flesh, juicy, dense, tender.

**Yield:** Average yield 20 t/ha.

**Disease and climate resistance:** Drought-resistant variety. Moderately diseased.

**Competitiveness:** The main advantages of the variety: Winter-hardy, regular fruiting, yield and stability of fruit formation in years with different meteorological conditions.

**Recommendations:** For cultivation in Almaty, Zhambyl and Turkestan regions (**Table 3**).

**Apple variety Egemen**

| Rootstocks | Tree height, m | Crown projection area, $m^2$ | Crown volume, $m^3$ | Average yield for the last 5 years, c/ha |
|---|---|---|---|---|
| Alma-Ata's region | | | | |
| Aport, 20-year-old trees | | | | |
| Niedzwiecki (St) | 4.6 | 14.0 | 28 | 26.2 |
| MM106 (St) | 4.0 | 9.6 | 16 | 29.2 |
| B16-20 | 3.4 | 7.6 | 12 | 37.8 |
| M 9 (St) | 3.4 | 9.2 | 17 | 93.6 |
| B7-35 | 3.4 | 8.8 | 15 | 129.6 |
| Dawn of Alatau, 20-year-old trees | | | | |
| Niedzwiecki (St) | 6.4 | 16.2 | 48 | 129.2 |
| MM106 (St) | 5.6 | 11.2 | 27 | 121.6 |
| B16-20 | 5.4 | 14.6 | 35 | 171.4 |
| M 9 (St) | 4.6 | 9.4 | 19 | 226.0 |
| B7-35 | 4.6 | 8.6 | 18 | 230.0 |
| South Kazakhstan region | | | | |
| Jonathan, 19-year-old trees | | | | |
| Niedzwiecki (St) | 4.0 | 12.6 | 22 | 75.6 |
| MM106 (St) | 3.6 | 10.2 | 15 | 107.2 |
| B16-20 | 3.2 | 11.0 | 16 | 162.0 |
| M 9 (St) | 2.9 | 9.0 | 9 | 168.0 |
| B7-35 | 2.8 | 8.2 | 9 | 160.0 |

**Table 3.**
*Influence of rootstocks on height, crown habit, and productivity of apple trees.*

**Year of inclusion of the variety in the State Register:** 2019
**Authors:** Vinovets A.D., Ostarkova L.V., Nurtazina N.Yu.
**Originator:** Kazakh Research Institute of Horticulture LLP
**Origin:** Derived from seedlings of the variety Golden Delicious.

**General characteristics:** Variety of winter ripening. The grade differs in high winter hardiness, plentiful annual fructification. Ripens at the end of September, lies until April. It begins fruiting 3–4 years after planting in the garden.

**Approbation signs:** The tree is medium-sized. The crown is dense, rounded, the branches are compact. The fruits are large, regular conical shape 180–200 g. With an integumentary okrus all over the fruit.

**Yield:** Average yield 20 t/ha.

**Disease and climatic resistance:** Drought-resistant variety. Resistant to powdery mildew and scab.

**Competitiveness:** The main advantages of the variety: Winter-hardy, regular fruiting, yield and stability of fruit formation in years with different meteorological conditions.

**Recommendations:** For cultivation in the Turkestan region.

**Apple variety Voskhod**

**Year of inclusion of the variety in the State Register:** 2011
**Authors:** Vinovets A.D., Ostarkova L.V.
**Originator:** Kazakh Research Institute of Horticulture LLP

**Origin:** Created by the method of intraspecific hybridization of individual selection from crossing Fantasia x Sinap Almaty.

**General characteristics:** Winter ripening, high winter hardiness. Resistant to powdery mildew and scab. It starts fruiting 2–3 years after planting in the garden. The yield is high.

**Approbation signs:** The tree is medium-sized, the crown is round, compact. The fruits are large, up to 260 g, candilla-shaped, light yellow in color with a delicate blush. The taste is sweet and sour, with a pleasant aroma. The pulp is white, dense, tender, juicy, fine-grained.

**Productivity:** Average productivity –35 t/ha.

**Disease and climate resistance:** Drought-resistant variety. Moderately diseased.

**Competitiveness:** The main advantages of the variety: Winter-hardy, regular fruiting, yield and stability of fruit formation in years with different meteorological conditions. Stored until April.

**Recommendations:** For cultivation in Almaty, Zhambyl and Turkestan regions.

**Pear variety Talgar beauty**

**Year of inclusion of the variety in the State Register:** 1965

**Authors:** Katseiko A.N.

**Originator:** Kazakh Research Institute of Horticulture LLP

**Origin:** Created by the method of intraspecific hybridization, the seedling of the Forest Beauty from St. pollination.

**General characteristics:** Autumn ripening. Winter hardiness is high.

**Approbation signs:** The tree is medium-sized. The crown is wide-pyramidal, of medium density. It enters fruiting in the 4th year after planting in the garden. Productivity is high. The fruits are large, elongated pear-shaped. The color is light yellow with a red carmine blush on the greater half of the fruit. The pulp is creamy, crispy, juicy.

**Yield:** Maximum yield 20.0 t/ha.

**Disease and climatic resistance:** Drought-resistant variety. Moderately diseased.

**Competitiveness:** The main advantages of the variety: Winter-hardy, regular fruiting, yield and stability of fruit formation in years with different meteorological conditions. Keep until January. Transportability is high.

**Recommendations:** For cultivation in Almaty, Zhambyl, East Kazakhstan, Kyzylorda, and Turkestan regions.

**Pear variety Fragrant**

**Year of inclusion of the variety in the State Register:** 1965

**Authors:** Katseiko A.N.

**Originator:** Kazakh Research Institute of Horticulture LLP

**Origin:** Created by the method of intraspecific hybridization, the seedling of the Forest Beauty from St. pollination.

**General characteristics:** Autumn ripening. It has high winter hardiness and good resistance to pests and diseases. It begins fruiting 5–6 years after planting in the garden. Yield is high.

**Approbation signs:** The tree is medium-sized. The crown is dense, pyramidal. The shape of the fruit is broadly pear-shaped, slightly oblique at the stalk. The color is yellow-green, with a slight blush on the sunny side, the flesh is tender, juicy, sweet with a strong specific aroma. Variety of winter ripening.

**Yield:** Maximum yield 18.0 t/ha.

**Disease and climate resistance:** Drought-resistant variety. Moderately diseased.

**Competitiveness:** The main advantages of the variety: Winter-hardy, regular fruiting, yield and stability of fruit formation in years with different meteorological conditions. Stored until January. Transportability is high.

**Recommendations:** For cultivation in Almaty, Kyzylorda regions.

**Plum variety Bailyk**

**Year of inclusion of the variety in the State Register:** 2020

**Authors:** Nurtazin M.T.

**Originator:** Kazakh Research Institute of Horticulture LLP

**Origin:** Created by intraspecific hybridization Renklod green from free pollination.

**General characteristics:** Autumn ripening, high winter hardiness. Disease resistant.

**Approbation signs:** The tree is medium-sized, the crown is compressed, reverse-middle shape. It starts fruiting 5–6 years after planting in the garden. The yield is high. Fruits are medium, oval, purple in color, pulp: orange, juicy, medium density, hormonally sour sweet taste.

**Yield:** Maximum yield 4.8 t/ha.

**Disease and climate resistance:** Drought-resistant variety. Moderately diseased.

**Competitiveness:** The main advantages of the variety: Winter-hardy, regular fruiting, yield and stability of fruit formation in years with different meteorological conditions.

**Recommendations:** For cultivation in the Almaty region.

**Plum variety Renklod Talgarsky**

**Year of inclusion of the variety in the State Register:** 2020

**Authors:** Nurtazin M.T.

**Originator:** Kazakh Research Institute of Horticulture LLP

**Origin:** from free pollination of the Edinburgh plum variety

**General Characteristics:** Medium early maturing variety than standard Stanley late maturing variety. Winter hardiness and resistance to diseases and pests is good.

**Approbation signs:** The tree is medium-sized, spherical with a spreading crown. It begins fruiting 4–5 years after planting in the garden. Mixed fruiting. Fruits are medium, round purple-brown. The pulp is yellow, juicy, cartilaginous, of good sweet and sour taste, it separates well from the stone. Variety of universal use.

**Yield:** Maximum yield 4.0 t/ha.

**Disease and climate resistance:** Drought-resistant variety. Moderately diseased.

**Competitiveness:** The main advantages of the variety: Winter-hardy, regular fruiting, yield and stability of fruit formation in years with different meteorological conditions.

**Recommendations:** For cultivation in the Almaty region.

**Cherry variety Aigerim**

**Year of inclusion of the variety in the State Register:** 1996

**Authors:** Nurtazin M.T.

**Originator:** Kazakh Research Institute of Horticulture LLP

**Origin:** Yellow Drogana from free pollination

**General characteristics:** A variety of medium-late ripening, high winter hardiness. Disease resistant high.

**Approbation features:** The tree is not large, compact, the crown is compressed, drooping, of medium density. It starts fruiting at the age of 5, after planting in the garden. The yield is high. The fruits are large, attractive, yellow in color, with a bright blush. The taste is sweet and sour, with a pleasant aroma. The pulp is yellow, dense, juicy, pleasant sweet and sour taste.

**Yield:** Maximum yield 20.5 t/ha.

**Disease and climate resistance:** Drought-resistant variety. Moderately diseased.

**Competitiveness:** The main advantages of the variety: Winter-hardy, regular fruiting, yield and stability of fruit formation in years with different meteorological conditions.

**Recommendations:** For cultivation in the Almaty region.

**Sweet cherry variety Lyazzat**

**Year of inclusion of the variety in the State Register:** 1999

**Authors:** Nurtazin M.T.

**Originator:** Kazakh Research Institute of Horticulture LLP

**Origin:** from free pollination of Drogana yellow

**General characteristics:** Late ripening, high winter hardiness. Disease resistant.

**Approbation signs:** The tree is medium-sized, the crown is round, spreading. It begins fruiting 5 years after planting in the garden. Yield is average. The fruits are large, dark red, obtuse-shaped. The taste is sweet and sour, with a pleasant aroma. The pulp is dark red, crackling, very dense, juicy.

**Yield:** Maximum yield 10.5 t/ha.

**Disease and climate resistance:** Drought-resistant variety. Moderately diseased.

**Competitiveness:** The main advantages of the variety: Winter-hardy, regular fruiting, yield and stability of fruit formation in years with different meteorological conditions.

**Recommendations:** For cultivation in the Zhambyl region.

**Tauly blackcurrant variety**

**Year of inclusion of the variety in the State Register:** 2020
**Authors:** Kadirsizova Zh.K. Egorova G.I.
**Originator:** Kazakh Research Institute of Horticulture LLP
**Origin:** Bred by selection from seedlings of the variety Minai Shmyrev.
**General characteristics:** A variety of medium ripening. The variety has an advantage in winter hardiness and resistance to powdery mildew in relation to the standard variety "Minay Shmyrev." There is a patent No. 572 dated 10/20/2015.
**Approbation signs:** The bush is medium-sized, semi-spreading. Shoots are medium, straight. Leaves are 3-lobed, large, dark green. The leaf blade is naked, matte, smooth, convex. Fruit raceme of medium length (5-8 cm). The berries are large, black, rounded, the skin is medium with a dry margin. The average weight of berries is 1.8 g, the maximum is 2.0 g. The taste of berries is sweet–sour (4.5 b) with aroma. Versatile berries.
**Productivity:** Productivity is 65.0–67.0c/hectare.
**Resistance to diseases and climatic conditions:** Moderately affected by diseases.
**Competitiveness:** The main advantages of the variety: Winter-hardy, external presentation, fruit size, regular fruiting, yield and stability of fruit formation in years with different meteorological conditions.
**Recommendations:** For cultivation in the Zhambyl region.

**Blackcurrant variety Gulzat**

**Year of inclusion of the variety in the State Register:** 2019.
**Authors:** Kadirsizova Zh.K., Egorova G.I.
**Originator:** Kazakh Research Institute of Horticulture LLP
**Origin:** Derived from free pollination of the currant variety Leskovitsa.
General characteristics: A variety of medium ripening. A variety of medium ripening, winter-hardy, drought-resistant, resistant to powdery mildew.
**Approbation signs:** The bush is medium-sized, semi-spreading. Shoots are straight, light brown. The leaves are 3-lobed, large, dark green. The leaf blade is naked, matte, wrinkled, with blunt short teeth. The berries are large, black, round, with a dry separation (1.7 g). The taste of berries – 4.5 points, with aroma.
**Productivity:** Productivity on the average 67–70c/hectare.
**Resistance to diseases and climatic conditions:** Moderately affected by diseases.
**Competitiveness:** The main advantages of the variety: Winter-hardy, regular fruiting, yield and stability of fruit formation in years with different meteorological conditions.
**Recommendations:** For cultivation in the Almaty region.
Under field conditions, variety samples are stored for 10 plants of each variety, which makes it possible to obtain a completely objective assessment based on the results of many years of study. The most valuable varieties and hybrids are preserved in vitro during cold storage [19].

## 4. Conclusion

Currently, the gene pool of fruit tree rootstocks in Kazakhstan is 39 forms, species and types.

The collection from various fruit growing regions of the world allows replenishing the agrobiodiversity of fruit trees in Kazakhstan and makes it possible to use their genetic and biological potential in the production of fruit seedlings and create modern gardens.

In the field gene bank of Kazakhstan, 140 technical varieties are preserved, of which: 9 varieties with increased winter hardiness, 12 – resistant to fungal diseases; 230 table varieties, of which 18 varieties of super-early and early ripening, 50 varieties of raisin, 28 varieties of selection of the Kazakh Research Institute of Fruit Growing and Viticulture were selected.

## Author details

Yuliya M. Yefremova[1*], Marina V. Urazayeva[2] and Saule Sh. Kazybaeva[3]

1 Leading Researcher from Laboratory of Agrotechnology and Nursery of Fruit and Berry Crops, LLP "Kazakh Fruit and Vegetable Grow Research Institute", Kazakhstan

2 Head of Laboratory of Agrotechnology and Nursery of Fruit and Berry Crops, LLP "Kazakh Fruit and Vegetable Grow Research Institute", Kazakhstan

3 Board for Science, LLP "Kazakh Fruit and Vegetable Grow Research Institute", Kazakhstan

*Address all correspondence to: ydyo@inbox.ru

## References

[1] Forecasts of socio-economic development of the Republic of Kazakhstan for 2021-2025 (Approved at RBC meeting dated May 4, 2020, Protocol No. 9). Available online: https://www.gov.kz/memleket/entities/economy/documents/details/

[2] Kaldybiev FT. State support for agricultural complex in rk. Materials of the Republican scientific-theoretical conference. Seifullin Readings - 11: Youth and Science. 2015;**1**(2): 146-149

[3] Isaev SI, Urazaeva MV. Comparative evaluation of apple tree seedlings grown on rootstocks of Russian selection in the conditions of the south-east of Kazakhstan. International Scientific and Practical Conference (July 22-23, 2016). Almaty: Kainar. 2016, pp. 222-224

[4] Dzhangaliev AD. Wild Apple Tree of Kazakhstan. Alma-Ata: Science; 1977. p. 281

[5] Egorov EA, Buntsevich LL. Technological foundations and organization of the production of seedlings and other planting material for horticultural crops. Horticulture and Viticulture. 2018;**2**:39-41

[6] Kulikov IM, Zavrazhnov AI, Upadyshev MT, Borisova AA, Tumaeva TA. Scientific and methodological foundations of industrial agro-technology for the production of certified planting material for fruit and berry crops in the Russian Federation. Horticulture and Viticulture. 2018;**1**:30-35. DOI: 10.25556/VSTISP.2018.1.10500

[7] Korolev EY, Krasova NG, Galasheva AM. Influence of separate methods of stimulating the branching of annual apple tree seedlings. Horticulture and Viticulture. 2018;**3**:42-47. DOI: 10.25556/VSTISP.2018.3.14173

[8] Kazybaeva SZ, Suyunbaeva GM. Agrobiological Assessment of Some Introduced Grape Varieties in the Conditions of the South-east of Kazakhstan. Status and Prospects of Horticulture in the Ural-Volga Region and Adjacent Territories. Russia: GNU Orenburg OSSV; 2013. pp. 126-130

[9] Beresneva LV, Kazybaeva SZh, Serdyukov YuG. Conservation and study of genetic resources of grapes in Kazakhstan. Proceedings of the International Conference "Mobilization and conservation of genetic resources of grapes, improvement of breeding process methods". Novocherkassk. 2008. S.33-S.36

[10] Rybakov AA, Ostroukhova SA, Gorbach VI. Viticulture: Proc. manual for students of page - x. in-comrade. 2nd ed. Tashkent: Ukituvchi; 1975. p. 136

[11] Lazarevsky MA. The Study of Grape Varieties. Rostov-on-Don: Publishing House of the Rostov University; 1963

[12] Guidelines for the establishment of experiments with fruit and berry crops and grapes in the Kazakh SSR. Proceedings of Kaz. NI-IPiV-Alma-Ata. Vol. 1, part 2. 1961

[13] Smirnov KV. Breeding large raisins and raisin grape varieties for the Uzbek SSR: Candidate's dissertation s.-x. Sciences/Fruit. in-t im. Michurinsk: I. V. Michurina; 1957. p. 166

[14] Kochanova I. Methods of studying clonal rootstocks in the Baltic republics and Belarus. Jelgava. 1980. p. 58

[15] Sedova EN, Ogoltsova TP, editors. Program and Methods of Variety Study

of Fruit, Berry and Nut Crops. VNIISPK: Eagle; 1999, 608

[16] Research Methods and Variation Statistics in Scientific Fruit Growing (Intern. Scientific and Methodological Conf.). - Michurinsk: MGSKhA, Vol. 3. 1998. p. 94

[17] Devyatov AS. Increasing the fertility of trees and productivity of orchards. In: Devyatov SA. 2nd ed. Minsk: Urajay. 1985. p. 216

[18] Karychev KG. Methodical instructions. In: Accelerated Assessment and Forecasting of Economic and Biological Characteristics in Studies with Fruit Crops. Kazakhstan: Almaty; 2001. p. 29

[19] Isaev SI, Urazaeva MV. New clone rootstocks in Kazakhstan. Scientific Works. Scientific support in the aspect of import substitution. Krasnodar. 2016; **10**:S.91-S.95